RESTORING AMERICA'S MILITARY PROWESS

RESTORING AMERICA'S MILITARY PROWESS

Creating Reliable Civil-Military Relations, Sound Campaign Planning, and Stability-Counter-insurgency Operations

John E. Peters

Foreword by Brigadier General John P. Rose
(U.S. Army, Retired)

ROWMAN & LITTLEFIELD
Lanham • Boulder • New York • London

Published by Rowman & Littlefield
A wholly owned subsidiary of The Rowman & Littlefield Publishing Group, Inc.
4501 Forbes Boulevard, Suite 200, Lanham, Maryland 20706
www.rowman.com

Unit A, Whitacre Mews, 26-34 Stannary Street, London SE11 4AB

British Library Cataloguing in Publication Information Available

Library of Congress Cataloging-in-Publication Data
Names: Peters, John E., 1947– author.
Title: Restoring America's military prowess : creating reliable civil-military relations, sound campaign planning, and stability-counter-insurgency operations / John E. Peters.
Other titles: Creating reliable civil-military relations, sound campaign planning, and stability-counter-insurgency operations
Description: Lanham, MD : Rowman & Littlefield, [2016] | Includes bibliographical references and index.
Identifiers: LCCN 2016027208 (print) | LCCN 2016028117 (ebook) | ISBN 9781442274716 (cloth : alk. paper) | ISBN 9781442274723 (electronic)
Subjects: LCSH: Military policy—United States. | Civil-military relations—United States. | Military planning—United States. | Counter-insurgency—United States. | Irregular warfare—United States.
Classification: LCC UA23 .P4678 2016 (print) | LCC UA23 (ebook) | DDC 355/.033573—dc23
LC record available at https://lccn.loc.gov/2016027208

∞™ The paper used in this publication meets the minimum requirements of American National Standard for Information Sciences—Permanence of Paper for Printed Library Materials, ANSI/NISO Z39.48-1992.

Printed in the United States of America

This work is dedicated to my beloved Annie
and to the memory of my parents.

Contents

Acronyms and Abbreviations

Acronym/ Abbreviation	Explanation
AEF	American Expeditionary Force
AOR	area of responsibility
AQAP	Al Qaeda in the Arabian Peninsula
AQ/VEO	Al Qaeda/violent extremist organization(s)
AVF	All Volunteer Force
BENELUX	Belgium, Netherlands, Luxembourg
CAPE	Cost Assessment and Program Evaluation
CBRNE	chemical, biological, radiological, nuclear, explosive
CENTCOM	Central Command
CIA	Central Intelligence Agency
CJCS	Chairman of the Joint Chiefs of Staff
COIN	counter-insurgency
CPM	Communist Party of Malaya
CRS	Congressional Research Service
CT	counter-terrorism (generally), communist terrorist in the context of Malaya
DCI	Defense Capabilities Initiative
DoD	Department of Defense
DRV	Democratic Republic of Vietnam
EFP	explosively formed penetrator (bomb enhancement)
EOD	explosive ordnance disposal
FLN	National Liberation Front (Algeria)

GEN	General
GOP	Grand Old Party
HADR	Humanitarian Assistance and Disaster Relief (operation)
HASC	House Armed Services Committee
HOA	Horn of Africa
IED	improvised explosive device
IFOR	implementation force
ISAF	International Security Assistance Force
JCS	Joint Chiefs of Staff
JOC	Joint Operations Center
JP	Joint Publication
JSOC	Joint Special Operations Command
JSOTF	Joint Special Operations Task Force
KAU	Kenya Africa Union
KFOR	Kosovo Force
KLA	Kosovo Liberation Army
KMT	Kuomintang (nationalist Chinese party)
LTG	Lieutenant General
MG	Major General
NATO	North Atlantic Treaty Organization
NAVEUR	(U.S.) Navy, Europe
NEO	Noncombatant Evacuation Operation
NSC	National Security Council
NSS	National Security Staff
OAF	Operation Allied Force
OAS	Organization of American States
OEF	Operation Enduring Freedom
OIF	Operation Iraqi Freedom
OND	Operation New Dawn
OOTW	Operations Other Than War
OPFOR	Opposition Force
OPLAN	Operations Plan
OSD	Office of the Secretary of Defense
POW	prisoner of war
PPBS	Planning, Programming, Budgeting System
Pub. L.	public law
RFID	radio frequency identification
ROE	rules of engagement
ROTC	Reserve Officers Training Corps
SACEUR	Supreme Allied Commander, Europe
SAMS	School of Advanced Military Studies
SFOD	Special Forces Operational Detachment

SIGAR	Special Inspector General for Afghanistan Reconstruction
SOFA	Status of Forces Agreement
UNOSOM	United Nations Operation Somalia
UNPROFOR	United Nations Protection Force
USAFE	U.S. Air Force, Europe
USAID	U.S. Agency for International Development
USCINCEUR	U.S. Commander in Chief, Europe
USSOUTHCOM	U.S. Southern Command
USSR	Union of Soviet Socialist Republics
WMD	weapons of mass destruction

Foreword

WHY DOES THE U.S. MILITARY, despite its $585.2 billion budget, all-volunteer force, and world-class equipment and facilities, produce such unsatisfying results? This is the question John E. Peters addresses in this short but important book. Peters acknowledges that the circumstances that lead to the use of military force are always complicated, and that second- and third-order effects of U.S. actions are often difficult to anticipate. Nonetheless, as he argues in this book, there are other systemic defects in the way the U.S. military and its commander-in-chief, the president of the United States, manage civil-military relations, undertake campaign planning, and prosecute counter-insurgency and stability operations that undermine the prospects for success. He offers recommendations to correct these defects.

The book's recommendations all require some level of congressional intervention, a circumstance many readers may find curious, but Peters argues effectively that Congress, because of its legislative and budgetary authorities, is the only actor in the federal government that can compel the president and the Department of Defense to undertake the necessary reforms.

Peters is highly qualified to write this book. His career includes twenty-four years in the U.S. Army with combat in Vietnam and Grenada, culminating with duty in the Army Staff's directorate of Strategy, Plans, and Policy. He subsequently spent more than twenty years at the RAND Corporation, where he worked on defense and international security issues for the U.S. Army, U.S. Air Force, U.S. Navy, U.S. Special Operations Command, and the Office of the Secretary of Defense. He has crafted a carefully thought-out analysis of important issues that undermine military performance. Future presidents, Congress, and the Department of Defense should take his book seriously.

Brigadier General John P. Rose, PhD (U.S. Army, Retired)
Bethesda, Maryland
April 2016

Preface

THIS BOOK IS A PRODUCT of my professional experience, including more than twenty years in the U.S. Army and another twenty years as an analyst at the RAND Corporation. My army service included combat in Vietnam and Grenada and culminated in the Strategy, Plans, and Policy directorate of the Army Staff. Much of my RAND work took place in support of the army, navy, and air force, as well as the special operations community. Over the combined forty-some years, it was my privilege to serve with many dedicated, hard-working service members of all ranks and, subsequently in my civilian capacity, to advise them and help them grapple with some of their most difficult problems. This book is meant to help them with a lingering and very profound problem: the inability to win the nation's wars, particularly those initiated in response to the September 11, 2001, terrorist attacks.

After devoting so much to the cause, it is hard to be objective about recent U.S. military operations. Reflection and self-criticism are difficult tasks for most people, and more difficult still for soldiers, Marines, sailors, and airmen who have spent many months away from their families and in harm's way, seen their comrades killed and wounded, and suffered the privations of the battlefield. Having invested so much, many believe that they did their best; if changes are needed, "they're above my pay grade." Others have told me "all our operations were successful." They are not alone in the desire to move on; the Department of Defense also has little stomach for detailed postmortems. I hope this book will provide useful insights that the Department of Defense might otherwise not discover.

This book focuses on a few issues that, in my judgment, constitute the root causes of the United States' recent military failures. As root causes of trouble, getting these issues right is imperative to future success, which makes them prerequisites that, if properly addressed, could help the U.S. military recover its warfighting prowess and restore the use of military force as a viable policy option for future presidents. Wars are dicey affairs, of course, in which the enemy and uncertainty also play major roles in the outcomes. However, a military that can satisfy the prerequisites discussed in this book will have a greater chance for success than one that does not.

The arguments presented here will probably anger senior military officials because they, like their more junior comrades in arms, have also invested heavily in America's recent wars and paid a heavy price for their service. Criticism—no matter how well intentioned—from an outsider who has never set foot in Afghanistan or Iraq is unlikely to be welcome. I hope that the ideas presented here will stand on their own merits and attract the attention of members of Congress. All of the changes advocated in these pages require congressional action in order to become reality.

Finally, this book is my own project, a work of the heart, undertaken in retirement. I have no remaining connection to the Department of Defense, RAND Corporation, or any of its sponsors. All views expressed and any errors are mine alone.

Santa Monica, California
September 2015

Acknowledgments

MANY PEOPLE HELPED BRING THIS WORK to fruition. First, my wife, Annie, encouraged me to take it on and let me transform our dining room into a study in which to write. My brothers, Darrell and David, were kind enough to act as readers of the earliest drafts of this work, offering constructive suggestions. They helped identify those instances when I had slipped unconsciously into military jargon unintelligible to the nonspecialist reader.

Friends and colleagues from within the U.S. Department of Defense and RAND also provided invaluable commentary on the manuscript. In this regard, I am indebted to Dr. Edward Keating and Dr. Elvira Loredo at RAND, and to Dr. Benjamin Zycher, a RAND alumnus now at the American Enterprise Institute, for their insights and helpful suggestions. Dr. Keith Dunn, former senior member of U.S. Mission NATO; Colonel Kurt B. Reineke (U.S. Army, Ret.); and Ms. Abigail Robinson, late of the Office of the Secretary of Defense (Policy) and currently part of the Defense Institution Reform Initiative, all were generous enough to read and comment on the manuscript, which is the better for their efforts. Finally, my thanks to Brigadier General John P. Rose (U.S. Army, Ret.) for his friendship and mentorship, and, in particular, for providing the foreword for this book.

1

Introduction

Most people reviewing the military outcomes in Iraq and Afghanistan would conclude that they were substantially less successful than Presidents Bush and Obama had hoped for. Combat operations in Iraq spanned 2003 to 2011 and in their wake left a shattered state riven by deep sectarian and tribal disputes and vulnerable to attack by the Islamic State and similar actors. Afghanistan, ravaged for a decade by the Soviet Union and under the thumb of the Taliban government when U.S. forces arrived in 2001, today verges on becoming an ungoverned territory as U.S. and coalition forces reduce their presence and Afghan forces step unsteadily into the breach. Meanwhile, al Qaeda and its fellow travelers bide their time, waiting until conditions are optimal for their assault on the fragile Afghan government.

How could this be? How is it that the U.S. military, self-described in 2001 as the "best trained, best equipped, and most capable in the world,"[1] could be fought to a draw by the irregular warriors and tribal militias that opposed it? After all, before sequestration, the United States out-spent the next fourteen states on the planet in terms of military budgets. Moreover, since the inception of the All Volunteer Force (AVF) in 1973, the Department of Defense (DoD) invested significantly in the professional development of its officers and noncommissioned officers, and in major training centers[2] where units could confront Opposition Forces (OPFOR) expert in enemy battlefield practices under simulated combat conditions. In addition, as the combat videos often found on YouTube and similar social media websites attest, U.S. forces were certainly brave under fire, and capable of profound destruction.

Nevertheless, despite the bravery of the troops, the tactical excellence of their leaders, and the high technology of their weapons and equipment, the United States has spent $1.6 trillion, lost 6,830 lives, suffered 52,310 wounded in action,[3] and has very little to show for its efforts.[4] The Taliban remains viable in Afghanistan and parts of Pakistan. Al Qaeda has survived U.S. efforts to eradicate it and has spawned franchise operations beyond the Middle East. U.S. military operations in Iraq contributed to circumstances that amplified tensions between Sunni and Shia that have plunged the country into a sectarian civil war. The nation's experience in Iraq and Afghanistan may be understood as a cautionary tale about war as an instrument of policy. The Obama administration seems to have understood the lesson to be that the military instrument is no longer a useful tool of policy; its employment presents too much uncertainty about the potential results, and it is too expensive in terms of blood and treasure.

What's the problem? It surely cannot be funding, quality manpower, sufficient equipment, or training facilities, given the size of DoD investments in these areas. The research for this book postulates that the problems bedeviling U.S. military effectiveness are conceptual and intellectual. The book will demonstrate that the military produced the outcomes witnessed in Afghanistan and Iraq in large part because of:

- poor civil-military relations
- an inability to formulate and execute sound campaign plans
- a mistaken approach to counter-insurgency and irregular warfare/stability operations
- inattention to military options other than regime change

While these issues are foremost in this author's judgment as major shortcomings in U.S. military operations, they also populate the pages of memoirs penned by former administration officials,[5] and the authors and veterans who have chronicled America's wars.[6] These issues are worthy of attention in part because they are systemic problems at the very foundation of U.S. military planning and operations. They exert profound effects on military outcomes no matter what the specifics of the geopolitical, global economic, or societal circumstances. As such, they cannot be overcome by the bravery and tactical proficiency of U.S. forces; untreated, they will likely undermine any future operations, or even deter a future president from considering military options for defending and advancing U.S. vital interests.

Civil-military relations focuses on the suitability of military advice rendered to the president by the senior uniformed officers, and the level and appropriateness of civilian involvement in military planning and operations. Civil-military relations capture the "nexus" of national policy and military force; if the splice between military and civilian responsibilities is flawed, then the United States may proceed (and has) with poorly conceived campaigns.

The ability to formulate and execute sound campaign plans is the sine qua non of military competence. It involves designing and sequencing a set of military activities in a way that will overcome the enemy's resistance, make it vulnerable to physical destruction, and compel the enemy leadership to capitulate to U.S. policy demands (e.g., abandon policies and activities the United States finds are contrary to its interests) or risk annihilation. Military action in the absence of solid campaign plans can be destructive, but also pointless, as the United States has experienced from Vietnam to Somalia to Afghanistan and Iraq.

A mistaken approach to counter-insurgency and irregular warfare/stability operations can leave the United States vulnerable to prolonged, indecisive campaigns. Counter-insurgency and stability operations are difficult under the best circumstances and near disastrous if not designed with appropriate population control and security measures, and resourced with force-to-population and force-to-space ratios appropriate to the local security challenges. Force size, however, while a prerequisite for success, is not sufficient on its own. Prevailing approaches to counter-insurgency and stability require that the U.S. forces deployed must have a clear understanding of the enemy and the local environment: things that recently have been beyond the reach of U.S. senior officials.[7]

Inattention to military options other than regime change may be the result of the United States' recent focus on Afghanistan and Iraq where the regime leadership seemed to lie at the heart of the threat these countries presented to U.S. national security. Even before 9/11, however, then-Chairman of the Joint Chiefs of Staff General Colin Powell was skeptical of limited objective operations.[8] Nevertheless, for most of its history, the United States' security needs have been met by less expansive (and risky) limited objective operations. It is time to reconsider their utility.

The book devotes a chapter to each of these topics. It concludes with a chapter offering advice to Congress on how to remedy these problems and as a result, restore U.S. military performance in future operations.

Sources and methods. The book is not intended as a detailed history of recent operations. It draws on published accounts of military and political events to provide context for analysis of the issue at hand (e.g., civil-military relations) and the effects of that issue on U.S. military performance. Readers seeking in-depth accounts of specific military actions should look to the works cited in the bibliography.

The research relied on unclassified sources, including the memoirs of George W. Bush, Donald Rumsfeld, Condoleezza Rice, Douglas J. Feith, Robert M. Gates, and others. Additional sources included official reports published by the DoD and internal DoD papers posted on various websites or published as appendixes in official reports or memoirs of the various officials involved. The selected bibliography contains full references.

The analytical approach in chapter 2 on civil-military relations presents a definition derived from the works of scholars on the subject. Next, the chapter summarizes the recent history of civil-military relations, focusing on military engagements from the end of the Cold War onward to today's operations. With the historical record summarized, the chapter evaluates the quality of civil-military relations found in the historical cases. Finally, the resulting analysis characterizes the recent cases of civil-military relations and their effects on military planning and operations. The chapter concludes with recommendations that might promote healthier future civil-military relations consistent with improved military outcomes.

Chapter 3 begins with the assertion, backed by historical evidence, that recent campaign planning has been faulty and has not delivered the outcomes that U.S. presidents sought when they authorized military operations. The heart of the chapter offers an explanation for the poor performance of recent campaign plans and suggests actions that might eventually provide a remedy.

Chapter 4 presents the case that U.S. counter-insurgency and stability operations have been flawed—indeed, that the doctrine that shaped their planning was unsuitable given the size of U.S. and coalition armed forces relative to the enemy, terrain, and populations of the countries where the United States deployed its forces. The narrative also considers aspects of the canonical British and French counter-insurgency cases and demonstrates why insights from these experiences were inappropriate for the United States. Next, the chapter presents empirically derived criteria for successful counter-insurgency and stability operations and describes why U.S. and coalition forces were unable to satisfy them.

Chapter 5 examines the historical performance of military operations short of regime change, a class of operations sometimes called limited objective operations. Using a DoD report that compiled all U.S. military operations from 1798 to 2014, the chapter characterizes operations carried out between 1950 and 2014 in terms of their type, frequency, location, and success to give the reader a sense of the recent U.S. military history and the efficacy of its operations. This analysis highlights the relative rarity of regime change as a military objective, but also shows that a minority of limited objective combat operations was successful, and suggests a revised approach.

Chapter 6 speaks to Congress, suggesting actions it should take if it agrees with the analysis presented and wants to implement this book's recommendations. It also allows the author to discuss insights that cropped up in the course of research for the book that bear on U.S. military effectiveness but are not related to any of the main themes examined in the previous chapters.

Notes

1. Department of Defense, Quadrennial Defense Review Report, September 30, 2001, 7.

2. The National Training Center at Fort Irwin, California, and the Joint Readiness Training Center at Fort Polk, Louisiana, represent two examples of such facilities.

3. Casualties summarized for Operation Enduring Freedom (operations in Afghanistan), Operation Iraqi Freedom (operations in Iraq), and Operation New Dawn (subsequent operations in Iraq) from the DoD casualty website (https://www.dmdc.osd.mil/dcas/pages/casualties_ond.xhtml).

4. One report indicates the United States' economic assistance to Afghanistan accomplished nothing. The U.S. Special Inspector General for Afghanistan Reconstruction says he is investigating the Pentagon's efforts to spark that country's economic development, which cost between $700 million and $800 million and "accomplished nothing." See file:///Users/johnpeters/Desktop/SIGAR:%20Pentagon's%20Economic%20Development%20in%20Afghanistan%20'Accomplished%20Nothing'%20%7C%20Defense%20News%20%7C%20defens.webarchive.

5. See Lieutenant General (U.S. Army, Ret.) Daniel Bolger, *Why We Lost: A General's Inside Account of the Iraq and Afghanistan Wars* (Boston: Edmond Dolan Books/Houghton Mifflin Harcourt, 2014), for a stark analysis.

6. For the Iraq war, see the list at Amazon.com: http://www.amazon.com/s/ref=nb_sb_ss_i_8_4?url=search-alias%3Dstripbooks&field-keywords=iraq+war+books&sprefix=Iraq%2Caps%2C317. For Afghanistan, see Amazon's bibliography: http://www.amazon.com/s/ref=nb_sb_ss_i_1_21?url=search-alias%3Dstripbooks&field-keywords=afghanistan+war+books&sprefix=afghanistan+war+books%2Cstripbooks%2C282&rh=n%3A283155%2Ck%3Aafghanistan+war+books.

7. The journal *Foreign Affairs* (July–August 2014) presents a series of articles ("what really happened") examining Cold War crises in which, despite a presence on the ground, U.S. officials still had only the vaguest, often mistaken, sense of the forces at work and their intentions within the crisis. It is not clear that the information revolution has mitigated this shortcoming.

8. Suzanne C. Nielson and Don M. Snider, eds., *American Civil-Military Relations: The Soldier and the State in a New Era* (Baltimore, MD: Johns Hopkins University Press, 2009), 15.

2

Civil-Military Relations

The late political scientist Samuel Huntington characterized civil-military relations as the "principal institutional component of military security policy" where "military security policy is the program of activities designed to minimize or neutralize efforts to weaken or destroy the nation by armed forces operating outside its institutional and territorial confines."[1] He identified three groups of issues as central to military security policy: (1) issues of size, recruitment, and support, and the proportion of state resources devoted to military needs; (2) organization, composition, equipment, and deployment of forces; and (3) issues of utilization of military forces.

Since Huntington, many scholars have weighed in on civil-military relations. Huntington's emphasis on civilian control remains a prominent theme in this ongoing discussion, but there are others, too.[2] Historian and former State Department official Dale Herspring's work focuses on the amount of conflict present in civil-military relations, and the degree to which a president's management and decision-making style affects the amount of tension present in the relationship.

The point of this chapter is to consider the level of cooperation present in civil-military relations and to determine whether cooperative relationships lead to better outcomes in military operations. Let us begin by specifying the details of civil-military relationships that are important to identifying shortcomings in them. The definition of civil-military relations used in this chapter is the interaction between the president and his defense and national security advisors, both civilian and military, and their resulting ability to:

- analyze a crisis or develop a contingency plan for a current or potential crisis and formulate an appropriate response;
- identify the vital interests of the United States engaged by the crisis;
- craft a plan to advance or defend those vital interests;
- identify clear end states and conditions for the outcome;
- present a clear chain of causality that describes how all the elements of state power, military and otherwise, will be brought to bear to create the desired outcome; and
- in particular, specify the chain of causality for military action.

The quality and form of civil-military relationships depends in part on the preferences of the president. Specific details of his/her preferences aside, the relationship should ideally provide for questions and dialogue among the parties so that military officials understand what the president wants and can explain to the president what they can accomplish given the available resources, and satisfy the president that the military actions proposed—in coordination with all the other instruments of U.S. power—will create the outcomes he/she desires. In addition, a healthy civil-military relationship should include an audit and oversight function to prevent potentially dangerous oversights and missteps.

The rest of this chapter summarizes the ebb and flow of civil-military relations from the end of the Cold War to the present. The summaries illuminate the ways in which healthy, robust relations sometimes lead to positive outcomes (although not always), and the ways in which poor civil-military relations undermine ongoing military operations and contribute to flawed and incomplete outcomes. The summaries also highlight how subtle the civil-military deficiencies can be and how they can undermine American military success and therefore fail to deliver the policy outcomes that prompted use of force in the first place.

The Recent History of Civil-Military Relations

The circumstances that drive tensions in civil-military relations today differ substantially from the forces at work during the Cold War. Much of the tension between the uniformed and civilian leadership during the Cold War had to do with their respective judgments on what sort of constraints the standoff with the Soviet Union imposed on regional wars like those in Korea and Indochina, and what sort of military posture contributed the most to deterrence and stability within the East-West confrontation.[3] The rules of the game within the newly created Department of Defense also imposed unanimity on the Joint Chiefs of Staff (JCS). They were required to speak with one voice when rendering military advice. Such advice reflected the compromises required to create it, and civilian officials in the Kennedy, Johnson, and subsequent administrations found the JCS's advice of little value. President Kennedy even brought General Maxwell Taylor into the White House as an alternative source of military judgment. The service chiefs and their staffs also confronted a new set of rules and procedures brought to the Pentagon by Secretary of Defense Robert McNamara and his civilian "whiz kid" analysts. The new Planning, Programming, Budgeting System (PPBS) confronted the uniformed leadership with new techniques and approaches that the service staffs struggled to understand and adopt.[4] Mastery of PPBS moved many key decisions on budgets, equipment, organization, and research and development from the military services to the domain of the new civilian analysts.

The questions driving today's tensions in civil-military relations have been about when, where, and how to employ the United States' armed forces. More specifically, should they be deployed solely in defense of American vital interests, or are there other circumstances such as the "duty to protect" notion, where U.S. forces would be deployed in instances like the collapse of Yugoslavia to prevent a government from brutalizing some segment of its population? Should U.S. forces be employed to prevent warring factions from hijacking humanitarian relief supplies, as in Somalia? In addition, there is the question concerning counter-insurgency and stability operations and how they are conducted—indeed, should the United States embark on such activities, given its poor record in recent attempts? Should the armed forces undertake "nation building," which involves extensive efforts to modernize a country's society, improve its infrastructure and government services, and encourage its leaders to embrace representative government, in shattered countries like Afghanistan? Is nation building even a viable policy option given the expense of such efforts? Should the United States deploy a relatively small force against some future enemy "X" with the expectation that "shock and awe" will cause the enemy's prompt collapse, or invest the build-up time to deploy a large, robust force capable of addressing virtually all credible contingencies?

The rules of the civil-military interaction are also different today. The chairman of the JCS no longer requires unanimity with the other chiefs in order to present advice. The chairman, not the joint chiefs, is now the official source of professional military advice. Moreover, as University of North Carolina at Chapel Hill professor Richard Kohn has written,[5] the uniformed leaders in the current era have been quick to seize the initiative and offer their judgments unbidden by their civilian bosses. Moreover, successive reforms (e.g., Goldwater-Nichols) have established clear lines of responsibility for the military services to raise, train, and equip forces, the combatant commands to employ military forces, and the chairman of the JCS and his staff to transmit (and transliterate) instructions from the commander-in-chief (the president) to the military forces. Other reforms in defense acquisition have produced clear civilian and military roles into the business of developing and acquiring new equipment.

Military Operations Under President George H. W. Bush[6]

According to Dale Herspring, President Bush's management style was well suited to the military. The president built a team of trusted advisors including Brent Scowcroft, the national security advisor; Richard Cheney, the secretary of defense; and General Colin Powell, who succeeded Admiral William Crowe on September 30, 1989, as the chairman of the JCS. They in turn seemed to share a general sense that matters that were policy in nature belonged to the civilians and those that were military in nature were the purview of the uniformed military. "Bush allowed the military professionals to handle operational matters, justify the military budget, and work out new force structures and military doctrines."[7] The team functioned well, though not without some internal frictions. Relations between Secretary of Defense Cheney and Chairman of the JCS Powell were "difficult." General Powell resented Cheney's direct interaction with the service chiefs and the secretary's efforts to minimize Powell's role as principal military advisor to the president.[8]

Panama

The first new challenge in international affairs to the Bush administration began in 1989 in Panama. General Manuel Antonio Noriega, once viewed favorably in Washington because of his assistance in U.S. counter-drug operations, engineered an accelerating series of provocations against U.S. personnel, including an attack on the U.S. embassy. As relations between Panama and Washington deteriorated, Secretary of Defense Cheney fired General Frederick Woerner, the U.S. Southern Command (USSOUTHCOM) commander, for his public criticism of the Bush administration, claiming it was responsible for creating a "policy vacuum" in the region that left the United States unprepared to deal with the deteriorating situation in Panama.[9] Then-Chairman of the JCS Admiral William Crowe was "livid" about the firing, because Cheney had not consulted him.[10] Crowe's displeasure ultimately had no effect on operations because the admiral retired.

Noriega subsequently nullified the national election that had been won by the opposition and installed his own government. President Bush declared it illegitimate.[11] Noriega successfully defeated a coup attempt in October and maintained his defiance of the United States and the Panamanian opposition. On December 15, the Panama National Assembly declared war on the United States, although Panama took no immediate military action following the declaration. General Maxwell R. Thurman, the new USSOUTHCOM commander, and Lieutenant General Carl Stiner, the XVIIIth Airborne Corps/ Joint Task Force South commander (the commanders who would plan and execute the United States' military response), had, with Powell's consent, begun planning a large-scale operation.[12] The plan evolved through several iterations and was ready when needed.[13]

The next day, December 16, saw two provocative acts. Four U.S. military officers, including Marine First Lieutenant Robert Paz, were on their way to dinner at the local Marriott hotel when their car was surrounded by angry Panamanian civilians and Panama Defense Force personnel. As the crowd closed in on them, the four officers fled in their car, and the Panamanians opened fire on them, killing Lieutenant Paz. In a separate incident, Panamanian officials detained and interrogated a U.S. Navy officer and his wife. The officer was badly beaten and his wife threatened with rape.

General Powell saw these events as Noriega "crossing a line" and said so to Secretary of Defense Cheney, who agreed with him. Powell briefed the service chiefs and then the president and the rest of the National Security Council (NSC) on the Thurman-Stiner plan and the deteriorating situation in Panama. Herspring's account notes that Cheney asked many questions during the briefing but did not alter any of the plans.[14] The resulting military action became Operation Just Cause and removed General Noriega from power.

The Panama case seems to present a picture of cooperative civil-military relations. Cheney and Powell managed the tension between them professionally and kept it from becoming a distraction to policy formulation and crisis management. President Bush wanted Panama to have a legitimate government that would, among other things, respect the treaties in place with the United States, stop money laundering for drug traffickers, and stop intimidating and mistreating U.S. personnel stationed in Panama. The decision to use force to oust Noriega reflected the confidence that the Thurman-Stiner plan enjoyed, the advocacy from the chairman of the JCS, and the support of the secretary of defense. Powell and Cheney brought the president and his other advisors on board using established NSC procedures. The operation itself removed Noriega and cleared the way for a legitimate government for Panama, which was the president's objective. And according to a BBC article in 2014, Panama has made progress in terms of the quality of its governance, attracting foreign direct investment, and openness, though it continues to struggle with corruption and drug trafficking.[15]

Despite its successes, Operation Just Cause displayed shortcomings in terms of post-combat stability operations. Fortunately, the Panamanian people and their legitimate officials, assisted by U.S. armed forces personnel, were able to reestablish order, distribute critical supplies and water, and return the area to something approaching normalcy. Nevertheless, fully cooperative civil-military relations and comprehensive planning should have discovered that the post-combat plan, codenamed Blind Logic, had not been approved by the joint staff and no major unit (e.g., XVIIIth Airborne Corps) had been formally tasked with its execution.[16] Operation Just Cause illustrates how campaign planning at the highest levels can become overly focused on the critical combat phases of operations and neglect the later but equally important stability and transfer of authority phases. This tendency to overlook so-called Phase IV operations would recur in later U.S. military operations and prove to be a persistent problem that the United States has yet to resolve. Solid civil-military relations, and particularly the civilian oversight of military planning, should identify such shortcomings and ensure they are remedied.

Iraq (1990–1991)

According to Herspring, the United States anticipated that Saddam Hussein might act against Kuwait. On July 24, 1990, General Powell called General Schwarzkopf, the CENTCOM commander, and directed him to "come up with a two-tiered response" that would (1) address a "minor border infraction" and (2) recommend how to stop a larger action and protect the region. Schwarzkopf observed there was little to stop Saddam: no ground forces were nearby. Undersecretary of Defense for Policy Paul Wolfowitz argued that the United States should send Saddam a signal to demonstrate Washington's resolve, but Powell counseled against the notion of symbolic military action without clear objectives.[17] Iraq invaded Kuwait on August 2 and quickly overran the Kuwaiti defenses, advancing to the Saudi border.

The Bush administration's initial objectives were to protect Saudi Arabia and its oil fields and to deter Saddam from further aggression. The United States also had particular concerns about Iraqi chemical and possible nuclear weapons. Saddam had used chemical weapons in the war with Iran, and there was evidence that Iraq was pursuing nuclear weapons. The president declared that the use of such weapons of mass destruction (WMD) against allied forces would lead the United States to remove Saddam Hussein from power.[18] Short of such a circumstance, the president was leery of invading Iraq, noting, "Had we gone the invasion route, the United States could conceivably still be an occupying power in a bitterly hostile land."[19]

At an NSC meeting on the day of the Iraqi invasion, General Schwarzkopf debuted the initial plan employing the immediately available forces. Its first phase involved retaliatory strikes by the U.S. Navy at Iraqi army, political, and economic targets. The second phase was deliberate action by a one hundred thousand to two hundred thousand strong joint force. At an August 4 NSC meeting in Camp David, the Bush national security team briefed a revised plan that required a seventeen-week build-up of forces to protect Saudi Arabia and subsequent offensive operations to compel Iraq to leave Kuwait. The initial deterrent force (Operation Desert Shield) would include between 200,000 and 250,000 personnel. The eventual offensive force would be twice that size and would require eight to ten months to deploy and prepare.[20] The Saudis agreed on August 6 to host U.S. military forces on their soil.[21]

Planning continued over the next several months. General Powell involved himself in the politics of the problem (e.g., would there be popular support for military action) until Cheney directed him to focus on military issues.[22]

Secretary of Defense Cheney sought military advice from sources beyond Chairman Powell. Cheney consulted the navy and air force for their views even though the service headquarters under Goldwater-Nichols had no war-fighting functions; their tasks were to raise, train, and equip military forces for employment by the combatant commanders (e.g., General Schwarzkopf as CENTCOM commander). Nevertheless, prompted by Cheney, the air staff was substantially involved in planning the air campaign, a job that should have fallen to Air Forces Central Command, the air force component command under CENTCOM, reflecting the distribution of responsibilities established by the Goldwater-Nichols Department of Defense Reorganization Act of 1986 (Pub. L. 99-433).

On August 17, President Bush authorized reserve call-ups, and planning continued. The services began competing for roles in operations, to little effect. CENTCOM prepared troop lists based on its ongoing mission analysis. General Powell approved the units requested.[23]

September proved eventful in terms of civil-military relations. Cheney fired the chief of staff of the air force, General Michael Dugan, for impolitic remarks he made to the press about the details of future air operations against Saddam Hussein. President Bush grew nervous about the lengthy preparations for war. On September 24, General Powell briefed the president on the options: go on the offensive or continue reliance on sanctions. The president wanted a briefing on what an offensive would look like. Schwarzkopf sent his CENTCOM briefing team to brief the president and his national security team. The ground phase of the plan proved a source of significant concern, "appalling" Cheney and Scowcroft. It appeared to be a single corps direct assault into the Iraqi defenses.[24] Powell went to Saudi Arabia to consult directly with Schwarzkopf and make refinements in the plan.[25]

In the meantime, Cheney enlisted the joint staff to draft alternative plans. The secretary of defense intended to light "a fire under the military," the uniformed leadership of which Cheney thought was foot-dragging.[26]

Ultimately, the president increased the size of the ground force package to ensure there would be enough offensive forces. He committed a heavy corps, as General Schwarzkopf had requested.[27] The U.S. Army Seventh Corps, based in U.S. European Command, deployed to Saudi Arabia with its heavy divisions.

The campaign itself was a curious thing from the standpoint of civil-military relations and sound campaign planning. The ground war concluded after a mere one hundred hours, before some of the president's stated objectives had been accomplished. President Bush made the decision to cease operations with the full support of Generals Schwarzkopf and Powell[28] before allied forces had seized or secured (or even located) the bulk of Iraq's WMD. The cease-fire also left the Iraqi Republican Guard largely intact and left the regime in firm control of the country.[29] Moreover, Washington did not establish the cease-fire terms; General Schwarzkopf met with the Iraqi delegates at Safwan airfield uninstructed and the general used his own discretion, based on his tactical judgment.[30] The Bush administration did not develop its own view of how the war should end, a view that should have been the product of civil-military dialogue. It did not establish the chain of causality that would render the Iraqi regime unable to use WMD against allied forces.[31] It did not develop criteria that, when achieved, would leave Saddam vulnerable to future coercion and control by the United States and prevent him from brutalizing his own people while suppressing the northern and southern insurrections in the aftermath of the Gulf War.

The conventional wisdom that civil-military relations in the Bush administration were healthy and robust is mistaken. Truly healthy civil-military relations would have created an atmosphere in which the president's advisors could have reminded him of the tasks that remained unfinished. In such an atmosphere, military officials could have noted that the threat posed by Saddam's WMD had not yet been addressed, and that his surviving air defenses still represented a threat to coalition air forces and their ability to impose no-fly zones in Iraqi airspace. The civil-military relations within the Bush national security team led directly to the problems encountered during and after Operation Desert Storm and laid the groundwork for the tragedy that became Operation Iraqi Freedom a decade later.

Military Operations Under President William J. Clinton

One need only Google "bill clinton and the military" to encounter dozens of sites representing President Bill Clinton's efforts to avoid conscription as a young man. Some of the sites infer deep moral failures associated with the way they believe the president avoided military service. According to Herspring (and my own experiences), many within the military held similarly dim views of the president. Part of the ill will resulted from the president's treatment of the military. In particular, the 1993 pay freeze the president instituted, despite the fact that military pay lagged 20 percent behind the private sector and some twenty thousand enlisted members were eligible for food stamps, led to the impression the president did not care about the welfare of service members. The president's lack of urgency in filling Department of Defense positions likewise struck some military officers as signs of dereliction of responsibility by the commander-in-chief.[32]

Moreover, the president's preference for ad hoc, unstructured decision-making and his focus on domestic policy issues rankled. Clinton rarely attended NSC meetings and took "only a peripheral role in issues such as Somalia, Bosnia, and Kosovo, (relying) on his chief of staff to keep him informed."[33]

Some officers found the president morally wanting. One characterized him as a "gay-loving," "pot-smoking," "draft-dodging," and "womanizing" commander-in-chief.[34] This circumstance undermined the civil-military relationship because many of the senior uniformed officers believed they could not trust the president.

Somalia

President and military dictator Siad Barre was driven from power in January 1991 and Somalia quickly descended into clan-based civil war. As *The United States Army in Somalia 1992–1994* describes it:

> This situation led to a struggle over food supplies with each clan raiding the storehouses and depots of the others. Coupled with a drought, these actions brought famine to hundreds of thousands of the nation's poor. Although private and volunteer relief organizations established refugee camps to try to prevent widespread deaths from starvation, they could not handle the massive amounts of aid and the requisite security structure that were needed. International relief organizations paid protection money to the warlords as they tried to distribute what donated food supplies did arrive. More often than not, such supplies never reached the hands of those who needed them but instead were confiscated by the warlords who distributed or sold them to enhance their own power and prestige. The general misery was only compounded by the brutality of the Somali clans

toward their rivals and the sporadic outbreaks of actual fighting. The most visible elements of the suffering-pictures of starving, fly-covered children-appeared nightly on American television screens. Fresh from its triumph in Operation DESERT STORM, the administration of President George H. W. Bush felt it could not ignore the situation, despite the obvious risks of intervening in a country still at war with itself.[35]

The Bush administration sent food aid in August 1992, an action that president-elect Clinton endorsed since his administration would inherit the issue. The Pentagon agreed with this policy and offered military forces if they were needed. Even General Powell, usually cautious about military intervention, was supportive, though he counseled that U.S. involvement should be brief, handing off responsibility to the United Nations as quickly as possible. On December 9, 1992, President Bush, with Clinton's support, dispatched the advance elements of a multi-national force to protect aid distribution. Eventually troops from twenty-one countries would participate. This military effort was transferred to the United Nations in May 1993 (UNOSOM) and the residual U.S. military element was reduced to five thousand troops.[36]

For a time, the militias and UN forces tolerated each other. But, as Lt. General Thomas Montgomery, deputy UN force command and the commander of U.S. forces in Somalia, explained,

> Aidid [the principal warlord contesting international intervention efforts] did not want the United Nations there. The approach that the United Nations took was to build from the bottom up . . . to start at the village council level and the regional level. Not working with the warlords. And this was threatening stuff. . . . Aidid didn't want any part of that.[37]

The relationship soured further when UN Secretary General Boutros-Ghali directed UN forces to pacify the area and disarm the militias. President Clinton was sympathetic, but the service chiefs saw "mission creep," a term used to characterize the unwise expansion of a military operation's goals. President Clinton replaced the seasoned U.S. Special Envoy to Somalia, Ambassador Robert Oakley, with retired Admiral Jonathan Howe, an administration insider, to work with the United Nations. Howe "had little political sense and knew nothing of Somalia"[38]—perhaps not the optimal choice, given the deepening crisis.

On June 5, militia attacks killed and wounded seventy-five Pakistani soldiers. The United Nations responded with a resolution for the "arrest and punishment of those responsible." Questions remain as to whether President Clinton was consulted on the substance of the resolution and whether he approved it. Within his administration, Clinton's key advisors were split. Ambassador Howe wanted to capture General Mohammed Aidid, the warlord responsible for the attacks. Secretary of State Warren Christopher, Peter Tarnoff, the undersecretary of state for political affairs, and Madeleine Albright, then-U.S. ambassador to the United Nations, supported his view; Secretary of Defense Leslie Aspin, General Powell, and CENTCOM commander General Joseph Hoar opposed it,[39] fearing such an effort would involve U.S. forces in lengthy, costly, and futile operations.

A mine attack on UN forces on August 21 led the service chiefs and General Powell to recommend deploying Special Forces Operational Detachment Delta and a contingent of Rangers. Anthony Lake, the national security advisor, and Secretary of State Christopher agreed. "The president did not weigh in."[40] It appears that Clinton was disengaged while Powell and Aspin made decisions for him. They attempted to manage mission creep as conditions deteriorated, but Clinton abdicated his responsibilities on guiding strategic choices: Do we stay or leave? If we stay, to what end? General Powell did tell Clinton to "get those Rangers out of Somalia" just before the October disaster in which a Delta and Ranger raid intended to capture a Somali warlord degenerated into a sustained, deadly firefight later memorialized in Mark Bowden's book, *Blackhawk Down*.[41]

Civil-military relations during the Somalia engagement were broken because President Clinton abdicated his responsibilities as commander-in-chief. His national security team was divided on the appropriate extent of U.S. involvement. In particular, they held very different views on the degree to which U.S. interests were congruent with those of the United Nations in Somalia. The president should have been the one to resolve these disputes and to set the U.S. course of action in Somalia. Instead, Secretary of Defense Aspin and General Powell did the best they could in the absence of presidential leadership.

In the wake of the "Blackhawk Down" debacle, Somalia prompted a change in U.S. policy. The Clinton administration became more cautious about the use of military force and took no action when Rwanda exploded.[42] According to historian David Milne, "This episode had a major impact on Clinton's willingness to deploy the nation's military in areas that [American architect of U.S. Cold War containment policy] George Kennan viewed as peripheral."[43] Moreover, President Clinton signed Presidential Decision Directive 25 on May 6, 1994, in part to reduce U.S. obligations for peace support operations and in part to strengthen the UN capability for them.[44]

Haiti

By the time the Clinton administration confronted Haiti, the president's national security team had experienced important changes. William Perry became secretary of defense and General John Shalikashvili (known informally as "Shali") became chairman of the JCS. Perry was confirmed unanimously and was highly regarded by the uniformed military leadership. General Shalikashvili took a broader view than his predecessor, General Powell, on the use of military forces, and expected that they could conduct operations other than war (OOTW) successfully.[45]

The United States has a long history of involvement with Haiti.[46] President Woodrow Wilson dispatched the Marines there after the assassination of the Haitian president. The ensuing U.S. occupation endured from 1915 to 1934.[47] The Clinton administration became engaged after Lieutenant General Raoul Cedras overthrew the only popularly elected president Haiti had ever had, Jean-Bertrand Aristide. The Clinton administration worked through the Organization of American States (OAS) to impose a trade embargo on Haiti and demanded that President Aristide be restored to office. General Cedras and President Aristide, under the auspices of the United States and OAS, agreed to a restoration of the latter's power. As the State Department's official history summarized events:

> The so-called "Governors Island Accord" signed by Aristide and Cedras on July 3 called for Aristide's return to Haiti by October 30, 1993, an amnesty for the coup leaders, assistance in modernizing the Haitian army, and the establishment of a new Haitian police force. The agreement provided for the suspension of U.N. sanctions once Aristide had assumed office in Haiti.[48]

The Governors Island Accord found the president's national security advisors divided on the wisdom of intervention in Haiti. National Security Advisor Anthony Lake favored it while the secretary of defense and the service chiefs opposed it,[49] fearing it would produce an extended need for U.S. forces. President Clinton's decision to dispatch troops does not appear to be driven by concerns about U.S. interests but as compensation to Haiti for promises he made and reneged on as a presidential candidate. Columnist Michael R. Gordon notes, "In the view of some officials, the White House commitment to the restoration of democracy is, in part, an effort to compensate for breaking the campaign pledge on Haitian refugees."[50]

> Despite indications that the Haitian military was backing away from the agreement, the United States dispatched the USS Harlan County with 200 U.S. and Canadian engineers and military police on board to prepare for the return of Aristide. On October 11 the ship was met at the pier in Port-au-Prince by a mob of Haitians, appearing to threaten violence. With the street battle in Mogadishu only a week past, the administration proved unwilling to risk casualties in Port-au-Prince. The ship pulled away the following day and returned to the United States, a significant setback for the Clinton administration.[51]

Following the *Harlan County* incident, the Clinton administration sought and secured UN Security Council Resolution 940, which authorized the use of force to restore democracy in Haiti. Planning for an invasion to drive Cedras from power proceeded through the summer, with Clinton approving the final plan during an NSC meeting on September 7. As the invasion force boarded its aircraft, a special delegation consisting of former president Jimmy Carter, Colin Powell, and Senator Sam Nunn visited General Cedras in Port-au-Prince to urge him to step down. Cedras capitulated only when he learned that the 82nd Airborne Division was en route.[52]

Although ultimately successful, the Haiti operation was a muddle from the perspective of civil-military relations. As the synopsis above suggests, the president's engagement on the matter may not have been driven by considerations of U.S. interests but by other emotions, perhaps guilt. There was no clear identification of military problems to be solved beyond restoring Aristide to office. Nevertheless, the coercive diplomacy of the Carter delegation visit along with the 82nd Airborne Division in the air achieved the limited objectives the administration set for itself.

Balkans (Bosnia, Dayton Accords, Kosovo)

Yugoslavia began unraveling in 1991 as major sub-regions (Slovenia, Croatia, Bosnia-Herzegovina) within the federation declared their independence. In multi-ethnic areas such as Bosnia, independence also brought violence as local Serbs, Muslim Bosniaks, and Croats clashed over territorial claims, human rights, and their respective status in the emerging independent states. A series of negotiations[53] sought unsuccessfully to broker partition settlements. Several European states deployed peacekeeping troops to the region under UN auspices in an attempt to reduce ethnic cleansing and provide sanctuaries where refugees could receive aid in safety.

According to Herspring, the Clinton administration had no clear plan. The service chiefs and Congress had no interest in a military intervention. The national security community was split, with the Pentagon advising the use of force only *in extremis* but the NSC and State Department advocating making the threat and use of force integral to U.S. policy. Clinton himself feared being drawn further into Bosnia. In an April 29, 1993, meeting with the service chiefs and his civilian advisors, the president reviewed the administration's options. It could lift the embargo on arms to the Muslims. It might try selective bombing, though General Powell had earlier advised it was unlikely to be effective. This so-called lift and strike policy was not approved by the European states with troops on the ground in the area, who feared the Serbs would take some of them hostage.[54]

U.S. policy continued to evolve until the summer of 1995,[55] when two events combined to galvanize the United States into action. The first was the July 11–13 massacre of eight thousand Bosnian Muslims by the Bosnian Serb Army of Republika Srpska at Srebrenica[56]; the second was the attack on Sarajevo on August 28, which prompted air strikes.[57] Curiously, the air campaign that followed and that led to the Dayton Accords was not coordinated with U.S. policy. This came about, according to Herspring, because relations between the U.S. military and civilians involved were very poor. In particular, Admiral Leighton W. Smith, commander, Allied Forces Southern Europe/U.S. Naval Forces Europe, clashed with Ambassador Richard Holbrooke over the usefulness of the limited bombing campaign.[58] Admiral Smith ordered Air Force General Mike Ryan, the commander, Allied Air Forces Southern Europe, "to have no contact with the negotiating team" led by Ambassador Holbrooke.[59]

Annex 1-A of the Dayton Accords specified the military arrangements for peace and stability in Bosnia. It provided for a multi-national Implementation Force (IFOR) comprised of North American Treaty Organization (NATO) and non-NATO countries that would replace the United Nations UNPROFOR.[60]

Even before IFOR was operational, there was a major dispute within the Clinton administration over the military's exact role. The civilian officials favored active policing of troublemakers and war criminals. Admiral Smith, now IFOR commander, was not interested and claimed IFOR lacked the authority to make such arrests.[61] The president called Admiral Smith "insubordinate" but left him in command, instead replacing SACEUR/USCINCEUR General George Joulwan with newly promoted four-star Wesley Clark.[62] Ironically, William Cohen, the successor to William Perry as secretary of defense and the official who actually appointed Clark to his new position, soon became concerned that General Clark was "selling out to the other side"—that is, the White House and State Department. General Hugh Shelton, now the chairman of the JCS, was sent to caution Clark.[63] Relations between General Clark and Secretary Cohen deteriorated to the point that the secretary of defense and General Shelton began managing Clark's itinerary for his Washington, DC, visits to prevent him from meeting officials they distrusted.[64]

Then there was Kosovo. It had long been an autonomous province within Serbia, and it held deep emotional connections for Serbs. Many of their most cherished monasteries and historically important sites lay within its boundaries. The ethnically Albanian population, once a minority, had grown to outnumber Serbs in the region and sought independence from Belgrade. The Dayton Peace Agreement did not cover Kosovo, and Serbia was becoming more draconian in its efforts to suppress the movement toward secession.

The Albanians had formed a militia called the Kosovo Liberation Army, or KLA (UCK in Albanian), which, by 1997, possessed some fifteen thousand to twenty thousand members.[65]

According to Herspring, Secretary of State Madeleine Albright was running the administration's foreign policy at this point while President Clinton was sidelined by the Monica Lewinsky affair. In May 1998, Slobodan Milosevic began systematic retaliation against the KLA, which included the massacre at Frekaz, Kosovo, where KLA leader Adem Jashari and sixty of his family members were killed.[66] The next month, General Clark consulted with CJCS General Shelton and Secretary of Defense Cohen on air power to make Milosevic back off. The JCS vice chairman, General Ralston, asked Clark, "What if air strikes don't deter?" Clark replied, "Something on the ground."[67]

Kosovo was not a priority for most of the U.S. military leadership. Their focus was Saddam Hussein and North Korea.[68] As a result, Clark worked up the campaign plan for Kosovo largely within his own headquarters.

The Serbian murder of forty-five unarmed Albanian Kosovars at Racak[69] prompted a NATO response. The alliance convened an emergency meeting at NATO headquarters to discuss possible courses of action. Clark and German General Klaus Naumann, head of the NATO Military Committee, met with Milosevic in Belgrade to no avail; Milosevic insisted on "the right to defend ourselves." France hosted the February 1999 Rambouillet conference, which sought a peaceful reconciliation of Albanian demands for independence with Serbian insistence upon its territorial integrity and non-interference by third-party military forces.[70] Secretary of Defense Cohen limited Clark's role in the conference to working on the military annex to the agreement. The Serbs refused to sign it, however, which was the proximate cause of Operation Allied Force (OAF).[71]

The White House was determined to proceed with the OAF bombing plan. The Service Chiefs did everything possible to minimize U.S. involvement in it. At the time, the United States was sponsoring the Defense Capabilities Initiative (DCI), which intended to stop the deterioration in European NATO members' military forces by negotiating new force structure guidelines with the allies. The chiefs argued Operation Allied Force was well within the capabilities of the European allies and insisted that they do their fair share.[72]

By the NATO Summit on April 23, 1999, Clinton had announced publicly, "I do not intend to put our troops in Kosovo to fight a war." The president was not alone in his view; public opinion did not support ground operations, either. The JCS chairman, General Shelton, did not have clear guidance regarding evolving Kosovo policy and could offer Clark little advice. Secretary Cohen cautioned Clark to say nothing at the summit about ground forces, noting, "We have to make this campaign work or we'll both be writing resumes." Shelton instructed Clark to give no more briefings on the campaign. Ultimately, Clark and Cohen had terrible relations, in part reflecting Clark's attempts to function as the allied commander in the absence of sound guidance from the administration, which was a product of Clinton's management and decision-making style.[73] As Herspring writes, "Clinton understood neither the complexity nor the importance of bureaucratic structures, and the military is probably the most highly bureaucratic structure in Washington."[74]

Civil-military relations in the Clinton administration were shaped by at least four important factors. The first and most influential was the president's limited engagement in and detachment from foreign policy issues. Clinton was eager to focus on domestic issues and to delegate foreign and defense policy to others. As Herspring noted, he rarely attended NSC meetings, leaving them to his chief of staff. The president certainly had feet of clay where the military was concerned, and he may have realized he lacked moral authority over the military, as his private remarks to General Shelton following the president's final NSC meeting suggested:

> Hugh, I know the last few years have put a tremendous strain on you based on my . . . activity. I know the principles for which you stand, and I know the values and character, that our men and women in uniform expect—and possess—and in truthfulness, I have not lived up to those values; and yet, you have stuck by me.[75]

A second important factor shaping civil-military relations on President Clinton's watch was the presence of civilian officials favoring assertive use of military force and threats of force (e.g., Madeleine Albright, Tony Lake, and Richard Holbrooke). To be sure, other administrations had their hawks, but in the absence of strong presidential involvement in the national security realm, the dynamic in the Clinton administration was arguably different. Influential officials supported causes (e.g., food aid for Somalia, reinstatement of a duly elected president in Haiti, and prevention of criminal acts against segments of the population by officials in the Balkans) rather than clearly identifying U.S. interests at stake in a crisis and the military actions necessary to defend or advance them.

A third factor was that, far from contesting the civilian hawks in the administration, three successive chairmen (Powell, Shalikashvili, and Shelton) served as active enablers of various parts of the administration's defense policies. Powell recommended dispatching SFOD Delta and Rangers to Somalia. General Shalikashvili oversaw the Haiti operation and was instrumental in turning the force that had launched as an invading army into one that landed in Port-au-Prince as a stability and peacekeeping force. General Shelton helped Secretary of Defense Cohen keep General Clark in check and turned the allied air campaign over Kosovo into a success. There were no conflicts between JCS chairmen and secretaries of defense during the Clinton years.

The difference in approaches to giving military advice by the chairman of the JCS who succeeded Colin Powell does not appear to have made much difference. General Shalikashvili and his successor, General Hugh Shelton, both possessed their own strong attributes as military leaders. As U.S. Naval War College Professor Lyle J. Goldstein observed of General Shalikashvili,

> No yes-man for the Clinton administration, he frequently injected caution into deliberations. His professional demeanor helped make American foreign policy more effective by improving the quality of military advice given to the president.[76]

General Shelton characterized his relationship with Secretary of Defense Cohen in contrast to the relationship he subsequently had with Secretary of Defense Rumsfeld:

> There are two kinds of relationships between a Chairman and Secretary of Defense. There was the kind I had with Bill Cohen, where we worked together and protected each other's flanks, and there was the McNamara-Rumsfeld model, based on deception, deceit, working political agendas, and trying to get the Joint Chiefs to support an action that might not be the right thing for the country but would work well for the President from a political standpoint.[77]

A final significant factor on civil-military relations during the Clinton years was the influence of UN and coalition partners on U.S. policy making. UN and allied concerns had more weight with some of the president's advisors than others. Certainly Secretary of State Warren Christopher and UN Ambassador Madeline Albright were among the most attentive to the United Nations, in particular. The dialogue between the president and his advisors that should have taken place given healthy civil-military relations apparently did not occur. As a result, Clinton never established clear priorities on the relative influence of U.S. interests, UN policies, and concerns held by allies. In the absence of such guidance, it was UN Secretary General Boutros-Ghali's directive to go after the Somali militias that set in motion the chain of events leading to the Task Force Ranger firefight and death of eighteen U.S. service members as they attempted to capture General Aidid. The constraints on U.S. policy options in the Balkans, beginning with Bosnia, were in large part the result of allied concerns for the safety of their troops on the ground. Perhaps ironically, Operation Allied Force in Kosovo prompted the most grumbling among U.S. Air Force targeting officers and planners about the limitations under which they labored but was nevertheless widely regarded as a success for the United States and its allies and partners—despite the fact that the cease-fire outcome required the continuing presence of the allied KFOR security and stability force in the country.

One can see how these four factors interacted to influence military operations over the course of the Clinton years. Somalia was a shambles, in large part because of the president's detachment, and because of Boutros-Ghali's insistence on pacification efforts aimed at the warlords and their militias. Haiti was inelegant but ultimately effective. The combination of the Carter-Powell-Nunn visit and the coercive power of the 82nd Airborne Division compelled Cedras to capitulate. Kosovo marked the most solid performance of civil-military relations on military contingency planning. Secretary of Defense Cohen and Chairman Shelton set appropriate constraints on General Clark that kept him in line with U.S. policy, maintained alliance consensus on the operation, and ultimately led to Milosevic's defeat.

Military Operations Under President George W. Bush

The arrival of George W. Bush's administration brought tension between Secretary of Defense Donald Rumsfeld and the uniformed leadership within the armed forces. The president's management style was corporate and hierarchical. He dealt "almost exclusively" with the heads of governmental departments and delegated decision-making to cabinet officials. According to Herspring, his approach invited conflict between departments, and when an important action was given to one department, it would freeze out others. These practices allowed Secretary of Defense Rumsfeld to dominate national security policy.[78]

Rumsfeld was a veteran secretary of defense, having held the job during the Gerald Ford administration. Secretary Rumsfeld thought he knew national security better than the service chiefs, whom he regarded as still stuck in Cold War–vintage military problems and operations.[79] In addition, he thought the military had enjoyed too much freedom during the Clinton era and that it was time to reassert civilian control. According to General Shelton, Rumsfeld, while meeting with Secretary Cohen on transition details,

> wanted Cohen's take on how to control the admirals and generals, who in Rumsfeld's opinion, had gained far too much power under Cohen's watch. Rumsfeld believed it was important for the Pentagon's civilian side to keep those dumb warmongers in check.[80]

Once sworn in, Secretary Rumsfeld began to assert his authority immediately. As Milne observed, "Rumsfeld and [his deputy] Paul Wolfowitz . . . learned a lesson that they applied to the second Iraq War: Colin Powell and similarly risk-averse generals had to be detached from decision-making."[81] Partly informed by this perspective, Rumsfeld was dismissive of Chairman of the JCS General Shelton's views. In his memoirs, Rumsfeld wrote, "the shock of 9/11 had not provoked much originality or imagination from the Chairman or his staff."[82] According to Herspring, Rumsfeld sought to interfere with the chairman's role as advisor to the president, ignored military studies that might have offered useful insights, and interviewed all candidates for promotion above two stars (an unusual level of involvement in military promotions for a secretary of defense). "The Service Chiefs did not exist for Rumsfeld."[83] The secretary of defense's disregard of the service chiefs was puzzling, given their significant statutory roles in raising, training, and equipping military forces. When Shelton's tenure ended, General Richard B. Myers succeeded him as chairman, and Marine Corps General Peter Pace became vice chairman. Both men, according to Herspring, were selected for their obedience.[84]

Afghanistan

With 9/11, "Osama bin Laden abruptly ended Rumsfeld's campaign to rein in the armed forces."[85] President Bush was eager for a prompt response to the attacks. Rumsfeld notes his own desire to prevent a subsequent attack in the United States. But the secretary of defense found there was "no existing war plan for Afghanistan. Steep budget cuts through the 1990s meant we had little intelligence on Afghanistan."[86] George Tenet, the director of the Central Intelligence Agency (CIA), suggested inserting CIA operatives to collect intelligence and support targeting. General Shelton presented three military response options to Secretary Rumsfeld. Option One was to strike at the enemy with cruise missiles. Option Two supplemented Option One with manned bombers striking targets for several days. Option Three consisted of Options One and Two plus "boots on the ground." The activities and objectives for ground forces were not specified.[87]

On September 21, 2001, Secretary of Defense Rumsfeld, General Shelton, General Myers, General Tommy Franks (the CENTCOM commander), and Major General Dell Dailey, commander of the Joint Special Operations Command (JSOC), met with President Bush at the White House to brief their current plans. U.S. special operations forces would link up with Northern Alliance units. The Northern Alliance was seen as a credible alternative to the Taliban and a lever with which to build indigenous support.[88] According to Rumsfeld, there was a strong, civil-military collaborative effort to form a broader strategy:

> The U.S. strategic theme should be aiding local peoples to free themselves of regimes that support terrorism. U.S. SOF and intelligence personnel should make allies of Afghanis, Iraqis, Lebanese, Sudanese and others who would use U.S. equipment, training, financial, military and humanitarian support to root out and attack the common enemies.[89]

Condoleezza Rice, the president's national security advisor, observed that planning for defense in this era required more than just pursuing and destroying terrorists but also addressing the failed states that were breeding grounds for terrorism. She came to see the importance of building stable democratic institutions. While such institutions were undeniably appealing, creating them was another thing. Rice noted the Department of State was poorly organized for its role in the overall effort, and it struggled to deploy senior Foreign Service officers to Afghanistan.[90]

According to President Bush's memoirs, General Franks, the CENTCOM commander, ultimately presented a four-phased war plan. Phase One would connect Special Forces with CIA teams that would link up with the Northern Alliance and other indigenous resistance forces and "clear the way for conventional forces to follow." Phase Two would be a massive air campaign to take out Taliban and al Qaeda targets and provide humanitarian air drops for the Afghan people. In Phase Three, U.S. and coalition forces would hunt down remaining enemy fighters. "Finally we would stabilize the country and help the Afghan people build a free society."[91] When, according to his memoirs, the president asked his senior staff, "who's going to run the country?" there was silence. The president directed that they ask Secretary of State Powell to "work on a plan for a transition to democracy."[92]

President Bush believed that "Afghanistan was the ultimate nation building mission. We had a strategic interest in helping the Afghan people build a free society."[93] According to the plan, Afghanistan was to be responsible for its own government. Hamid Karzai, a politically influential former leader of the Popalzai tribe, was selected at the International Conference on Afghanistan in December 2001 to head the interim administration, with broad-based transitional authority for six months. A constitution was to be adopted eighteen months thereafter.[94] According to Herspring, the original plan for nation building was to bring in a UN protective force and then leave.[95] But the plan, which drove the U.S. position during the Bonn Conference in November 2001, stumbled from the outset. According to Douglas Feith, undersecretary of defense for policy, "we learned that the U.S. government lacked the tools to promote new leadership in Afghanistan as effectively as we had removed the old leadership."[96]

Other shortcomings soon became apparent. As Feith noted, the U.S. government was not well organized for "reconstructing the Afghanistan community after regime change. . . . Our armed forces and civilian officials lacked the institutions, authorities, and resources needed for reconstruction work."[97] "The Bush administration never put together a comprehensive strategy to counter ideological support for Islamist extremism."[98] Secretary Rumsfeld commented in his own memoirs, "If some later contended that we never had a plan for full-fledged nation building or that we under-resourced such a plan, they were certainly correct."[99]

It is this sort of strategy-resources mismatch that sound civil-military relations should prevent, or at least minimize. The military should be able to speak candidly about what elements in the strategy it can and cannot execute and explain why. The other departments within the executive branch (e.g., State Department) must be equally candid. Had the Bush administration realized the limitations and shortcomings of its assistance that began to appear after the rout of the Taliban and dispersal of al Qaeda during the initial planning for operations, it might have pursued a more modest strategy in keeping with the capabilities at its disposal. In the event, however, it pursued a nation-building strategy it lacked the manpower, skills, and other resources to implement success fully.

Iraq (2003–2011)

President Clinton signed the 1998 Iraqi Liberation Act into law, making the removal of Saddam Hussein a matter of U.S. policy. By 2001, according to Undersecretary Feith, international sanctions, the Iraqi Survey Group to monitor Iraqi WMD efforts, and the oil for food program "were all corrupted"; Saddam was not "in a box."[100] At a September 29, 2001, meeting in Rumsfeld's office, the secretary of defense tasked JCS Chairman Myers to begin preparing military options for Iraq. For regime change, Rumsfeld asked for an option "that would take only one or two months and 250,000 troops, not necessarily all from the United States"[101]—an odd tasking, starting with the secretary's constraints and excluding details of the objective besides "regime change." In January 2002, Rumsfeld and Franks began detailed planning, but without the involvement of the service chiefs.[102] Excluding the service chiefs was a potentially serious error. They are responsible for raising, training, and equipping military forces for employment by the combatant commanders. No other senior defense officials are as intimately aware of the capabilities and limitations of the current forces as the service chiefs. Without their involvement in planning future operations, combatant commanders may be unaware of recently developed capabilities and therefore not include them in their requests for forces. They may also be unfamiliar with the latest technical upgrades to familiar weapons and, as a result, miss opportunities to exploit their enhanced capabilities. Colin Powell, by then the secretary of state, warned that war could destabilize the entire region.

Rumsfeld made a serious effort to identify events that might complicate or undermine U.S. military efforts in Iraq. He dubbed the list of these events the "Parade of Horribles." Events on the list included failure to find WMD, failure to manage post-Saddam Iraq, an extended (ten-year) post-war effort, ethnic strife, and a negative world reaction to preemption.[103] As the coalition campaign to remove Saddam Hussein's regime unfolded, all of the items on the "Parade of Horribles" list became factors challenging the Bush administration and its military commanders in Iraq. Unfortunately, although Rumsfeld had identified these potentially important factors, he did not direct contingency planning to control or counter them.

The principals (department secretaries) on August 14, 2002, approved a document titled "Iraq: Goals, Objectives, Strategy." President Bush also approved it later that month. The paper's goals were recited in all JCS and CENTCOM documents, citing the original NSC paper. According to Feith, "this was the essence of the civilian leadership's broad instruction to the military commanders."[104]

Despite explicit guidance from Chairman Myers to CENTCOM "to refine planning for Phase IV operations," General Franks resisted, preferring to focus his attention on the war-fighting phases rather than on stability in Phase IV.[105] General Myers told Franks that he

> would be responsible for all aspects of governing in Iraq. The military will play a pivotal role in setting the conditions for accomplishment of strategic objectives for post-war Iraq. To ensure rapid success during the transition to Phase IV post hostilities operations, detailed coordination must begin to integrate interagency efforts at the earliest opportunity.[106]

General Myers and Secretary of Defense Rumsfeld ultimately realized the limits on Franks's Phase IV planning and on January 10, 2003, tasked Joint Forces Command to create a joint task force "to support CENTCOM's post-war planning efforts."[107] In other words, despite the conventional wisdom about Rumsfeld's ignoring of the service chiefs and practice of intimidating his uniformed subordinates, the secretary of defense was unable to compel his CENTCOM commander to do a thorough job of Phase IV planning for Iraq: not a sign of healthy civil-military relations.

When it was time to deploy forces to CENTCOM for Operation Iraqi Freedom, Rumsfeld screened the deployment orders and destroyed their logical sequence.[108] This action produced what might be called procedural surprise for the joint staff and the military services, which had to ready units for deployment. During Operation Enduring Freedom in Afghanistan, Rumsfeld had been content to use the official Joint Operational Planning and Executing System[109] tools to alert, stage, load, and transport units. His insistence on micro-managing which units actually deployed and the order in which they did so deprived CENTCOM of the forces it needed to assert coalition authority over Iraq and its population once the fighting ceased. The force deployed was too small, given the size and distribution of the Iraqi population, and lacked military police and other specialists for area and route security. Rumsfeld offered a rationale for his actions, noting the Time Phased Force and Deployment Data System operated like an on/off switch rather than as a rheostat to adjust the flow of forces to complement diplomatic efforts. The mass deployment of forces "might rattle allies, undermine diplomacy, and deploy troops just before Christmas and New Years."[110]

The secretary's argument is unconvincing. First, recall that he sought a plan that could be executed quickly in "one or two months"; he was not inclined to lose momentum once events began to play out. Second, the administration already had the support of forty-eight nations.[111] Having witnessed Operation Enduring Freedom in Afghanistan, they undoubtedly knew what they were signing up for and therefore would not have been "rattled" by the scene of massive U.S. force deployments. Third, the holidays were not a major consideration influencing troop deployments. The administration was clearly eager for action against Saddam. Moreover, the administration had shown no such qualms about deploying forces to Afghanistan immediately before the holidays (the battle of Tora Bora took place in December 2001). Finally, diplomacy had already run its course. The die was cast for military action when Iraq delivered its final unsatisfactory WMD disclosure package.[112]

Civil-military relations in the run-up to the Iraqi war were not helpful to the president's objectives. On one hand, Secretary of Defense Rumsfeld abandoned many of the established procedures within the Pentagon, excluding key leaders from planning, and micro-managing force requests and troop deployment orders. On the other hand, for some reason, he concluded that he could not share his skepticism with the president about the campaign's ultimate objective—bestowing democracy on the Iraqi people.[113]

Military Operations Under President Barack Obama

The Obama administration arrived in the White House confronting the Great Recession of 2008 as well as the ongoing wars in Afghanistan and Iraq. Candidate Obama had begun making distinctions between a war of choice—that in Iraq—and a war of necessity—coalition operations in Afghanistan—on the campaign trail before winning the presidency. President Obama used his February 24, 2009, speech to a joint session of Congress (in other years the State of the Union address) to focus solely on remedies to the recession, emphasizing employment, education, energy, and debt reduction. When his administration published its first National Security Strategy in 2010, it noted:

> Our Armed Forces will always be a cornerstone of our security, but they must be complemented. Our security also depends upon diplomats . . . development experts . . . and intelligence and law enforcement.[114]

The document continued, noting that the United States must modernize its alliances "to meet the challenges of a new century," and that "we will build new and deeper partnerships in every region."[115]

Viewed in contrast to his predecessor's strategy, the Obama administration appeared to be de-emphasizing the military dimension of national security policy, emphasizing the domestic sources of power, and calling renewed attention to diplomacy, intelligence, and "building cooperation with other 21st century centers of influence"[116] as key avenues to advance and defend U.S. interests around the world.

Robert M. Gates headed the new president's national security team as secretary of defense. A GOP loyalist, Gates had been asked by the president to stay on in his post from his time in the George W. Bush administration. Admiral Michael Mullen served as chairman of the JCS.

For the Obama administration, all of the tests of civil-military relations took place in the context of the ongoing wars. According to Rick Brennan, a civilian advisor to senior military commanders in Iraq between 2006 and 2011, the tension between the administration and the military was the "greatest since Carter." The president consistently rejected military advice in favor of that from his civilian inner circle. Communications between the military and the civilian leadership stopped short of the president. Perhaps most damning of all, the president was widely perceived as squandering the sacrifices made by the military over the years in Iraq.[117]

Withdrawal from Iraq

The U.S. military withdrawal from Iraq was a major source of civil-military tension. In a nutshell, most knowledgeable military officials in Iraq believed a residual U.S. military presence after the 2011 withdrawal date was essential to preserve the fragile progress that had been achieved, prevent another slide into sectarian violence, minimize the malign influence of Iran, and prepare the Iraqi security forces to defend the country against foreign attacks.[118]

According to Brennan, the military had been warning the administration of the looming problems that would attend a complete military withdrawal from Iraq:

> By late 2009, General Raymond Odierno concluded that the goals of U.S. policy in Iraq could not be achieved by the end of 2011. He shared this assessment with officials at U.S. Central Command and the Pentagon and with the staff of the National Security Council. . . . Despite the efforts of Odierno and others . . . a gap had opened up between the strategic goals articulated by the Obama administration and the resources and time the White House was willing to commit to achieving them.[119]

The basic problem confronting the U.S. forces in Iraq was that there was too much to do and too little time to do it in before the 2011 troop withdrawal deadline. At the same time, the Shia-dominated government of Iraqi President Nouri al-Maliki was taking aggressive steps to marginalize the Sunni minority, and by the 2010 national elections, Iraq found itself in a state of political paralysis. Sunni-controlled areas of the country had the potential to become safe havens for al Qaeda in Iraq (soon to morph into ISIS/Islamic State). Shiite militias armed, trained, and equipped by Iran were aligning with Shiite political parties and "constituted a shadow government of sorts." An in-country review concluded, "If U.S. forces completely withdrew by the end of 2011, it would be very difficult for the Iraqis to maintain the fragile gains made since 2007."[120]

U.S. military planners estimated they would need twenty-four thousand troops for the residual force: enough to counter Iranian influence, have a moderating effect on Maliki's sectarian inclinations, and support border and airspace control. Secretary of Defense Gates, anticipating the president's rejection of a large stay-behind force, directed the U.S. commander, General Lloyd Austin, to prepare options of less than twenty thousand personnel. General Austin and his staff developed force packages in the eight thousand to ten thousand range and descriptions of the missions these forces would be able to accomplish. Subsequently in August 2011, President Obama authorized Ambassador Jeffrey to begin negotiations to keep five thousand U.S. troops in Iraq.[121]

"Obama rejected the military's call for a large-scale presence to continue training the Iraqi army and to secure the Arab-Kurd border area near Kirkuk." In addition, the president set two conditions the Iraqi government had to satisfy before there could be a residual U.S. force after 2011. First, the president insisted on a formal parliamentary request for U.S. forces to remain in Iraq. Second, he required a Status of Forces Agreement (SOFA) that would grant U.S. personnel immunity under Iraqi laws. The Iraqi parliament made the necessary request for forces in October but declined to grant U.S. forces immunity.[122]

Brennan holds the view that the president negotiated in bad faith, knowing that the Iraqi parliament was likely to reject a demand for a SOFA, and using that as an excuse to terminate the U.S. military mission in Iraq, despite the fact that U.S. military personnel had been protected in the absence of a SOFA by a diplomatic note signed by the foreign minister.[123] The president simply wanted out. The concerns raised by the military were beside the point. It was the president's prerogative to ignore his military advisors and their recommendations, but the subsequent appearance of the Islamic State/Da'ash terrorist organization and its advances in Iraq suggest the president may have made a very poor choice. After all, ungoverned territory like large patches of Iraq have long been a U.S. concern because such areas can become havens for terrorists.[124]

2009 Afghanistan Additional Troop Request

Afghanistan generated its own share of civil-military tensions. The Bush administration had become suspicious of the military's request in 2008 for thirty thousand additional troops for the summer fighting season. As Gates notes in his memoir, the administration "thought [Admiral] Mullen [JCS chairman] and [General] Petraeus were trying to force the administration into a major decision" though Gates believed they had no ulterior motives.[125] The Obama administration was similarly suspicious of the 2009 troop requests, viewing them as a "squeeze play" that would alienate the Democratic Party base if approved, while the Republicans would "just play politics" on the matter of the request no matter what the administration did.[126]

Subsequently, when the military requested five thousand "enablers" (air ambulances, counter-improvised explosive device technicians, etc.), the president pushed back, demanding to know what had changed—was this mission creep? Secretary of Defense Gates observed after the incident, "I was deeply uneasy with the Obama White House's lack of appreciation—from the top down—of the uncertainties and inherent unpredictability of war."[127]

Finally, in November 2009 when President Obama made the decision for a surge in Afghanistan, he made Gates, Admiral Mullen, General Petraeus, and General Stanley McChrystal, the International Security Assistance Force (ISAF) commander in Afghanistan, swear to support his decision. But, as Gates writes, it was the vice president and the national security staff the president should have worried about:

> From the moment the president left West Point (where he had made the announcement) (Biden and the [National Security Staff (NSS)]) worked to show he had been wrong, that the Pentagon was not following his direction, and that the war on the ground was going from bad to worse.[128] The more our commanders touted any success in the field, the more the NSS looked for evidence they were wrong.[129]

The Military and the Press

As Gates described it, "During September [2009], several events fractured what little trust remained between the senior military and the president and his staff." General Petraeus gave an interview to the *Washington Post* in which he indicated there was no guarantee that more troops would bring success in Afghanistan, adding, "it won't work out if we don't." Then Admiral Mullen argued for more troops in Afghanistan during his confirmation hearing for his second term as chairman in a way "implicitly critical of the vice president's views." The president was said to be "livid" with his remarks.[130]

Later that month, a piece in the *Washington Post* by Bob Woodward quoted a leaked copy of the McChrystal counter-terrorism report: "McChrystal: More Forces or Mission Failure." General McChrystal subsequently appeared on the CBS news program *Sixty Minutes*, "in which he spelled out in detail how bad he had found the situation in Afghanistan and what needed to be done."[131]

The president was upset, according to Gates. Obama could not understand what was wrong: Were Petraeus, McChrystal, and Mullen trying to box him in? "What is wrong? Is it the process?" "Are they suspicious of my politics?" "Do they resent that I never served in the military?" Gates defended his subordinates to the president, characterizing McChrystal as naïve, having never served before at such politically sensitive levels. Gates told the president that Petraeus and Mullen were loyal but compelled ethically to offer their views in testimony and interviews.[132]

January 2011 Strategy Review for Afghanistan

According to the Congressional Research Service,

> By early 2011, senior leaders, including the Commander of NATO's International Security Assistance Force (ISAF), General David Petraeus, were pointing to discrete progress on the ground, though noting that such progress was still "fragile and reversible."
>
> In late 2010, NATO and the Afghan government agreed to pursue a key medium-term goal: the transition of lead responsibility for security to Afghans throughout the country by the end of 2014. The U.S. government has stated its intention to begin drawing down some U.S. forces from Afghanistan in July 2011, and also to maintain a long-term strategic partnership with Afghanistan beyond 2014.[133]

A key point of debate during the January strategy review was the pace of withdrawal from Afghanistan. General Petraeus, basically lobbying for a slow rate of withdrawal, did a series of interviews describing "gains that had been made in routing the Taliban from traditional strongholds in the south and in training Afghan troops."[134] At the same time, military estimates on the size of necessary Afghanistan force levels ranged between 352,000 and 378,000.[135] The military planners making the estimates were trying to answer a military-technical question: How many troops will be needed given U.S. assumptions about the quality of security and stability in the country? The civilian staff on the president's NSS, however, viewed the estimates through the lens of their own political concerns and concluded that the military was out to "jam" the president by calling for force levels inconsistent with his withdrawal plans.

Yemen

At this writing (Fall 2015) the fate of Yemen and the U.S. counter-terrorism strategy there is still unfolding. The legitimate government has fled the country, factional warfare has forced the United States to withdraw its special operations task force, and a Saudi-led coalition has been carrying out air strikes. Some additional signs of civil-military discord within the Obama administration, spawned by events in Yemen, are visible. Civil-military tensions appear around two main issues: first, whether the United States should oppose Iranian-backed Houthi militia progress in the country, and second, if there is no functioning government with which to partner in a "by, with and through" campaign[136] against al Qaeda in the Arabian Peninsula (AQAP), what military options remain viable for the United States?

Until recently, the Obama administration held up U.S. military cooperation with Yemeni forces as an example of successful counter-terrorism operations.[137] In 2014, however, Shia Houthi militia members swept into Sana'a, the capital, and President Hadi fled to Aden, the major coastal city. Eventually, he resigned his office on January 15. In February 2015, the U.S. Department of State closed the embassy. On March 21, the last U.S. special operations forces evacuated the country. These events rendered previous U.S. policy ineffective. As General Lloyd J. Austin, the CENTCOM commander, told Congress,

> The economy is in a steady state of deterioration for some time. . . . The lack of central government leadership coupled with Houthi expansion (and the evacuation of all U.S. personnel in February 2015) have made it exceedingly difficult for us to [conduct partnered or unilateral counter-terrorism (CT) operations] against AQAP. The Yemeni government has generally curtailed its CT operations, and this has allowed AQAP to regain some of its former territory and increase operations against government and security forces.
>
> If the situation continues to erode and it remains on a negative trajectory now, Yemen could fracture and we could end up losing a key partner in our counter-terrorism fight and cede additional ungoverned space to our adversaries to operate out of.[138]

A key question is whether the Houthis are part of the problem or part of the solution. The civilian officials in the Obama administration tend to view the Shia Houthis and the Sunni AQAP as "two sides of the same coin" and not good for the Yemeni people.[139] Knowledgeable military officers, however, view the Houthi militias as a potential asset:

> Military sources said that a number of regional special forces officers and officers at U.S. Special Operations Command argued strenuously against supporting the Saudi-led intervention because the target of the intervention, the Shia Houthi movement—which has taken over much of Yemen and which Riyadh accuses of being a proxy for Tehran—has been an effective counter to al Qaeda.[140]

A source quoted in the article, the Jamestown Foundation's Michael Horton, noted, "These constant reports that the Houthis are working for Iran are nonsense." Horton observed that the Houthis don't need Iran for arms because they have plenty, and they don't need Iranian training, either, because they've been fighting since 2012. Horton closed his commentary with a question: "Why are we fighting a movement that's fighting al Qaeda?"[141]

If the Obama administration concludes that it cannot make common cause with the Houthis in Yemen, there are no obvious alternative local factions that would make good partners for the United States: no one with whom the armed forces could work "by, with and through." And this circumstance may be the second source of civil-military friction. CENTCOM is fully committed to "support a whole of government approach to developments in Yemen, preventing Yemen from becoming an ungoverned space for AQ/VEO (violent extremist organizations)."[142] The command, in other words, expects significant contributions from the State Department, the U.S. Agency for International Development, and other actors within the executive branch of government for any future counter-terrorism effort in Yemen.

The second question with potential civil-military relations import is really one of strategy. What is it the administration expects to accomplish in Yemen? What is the military role in the overall endeavor? What is the chain of causality? What is it the administration expects the armed forces to do, and how does accomplishing those things contribute to producing the wider outcomes the White House would like to see? And, finally, how does the military leadership view these tasks? Can they be accomplished, given the available resources? If not, what alternatives might they propose?

Conclusions and Observations

Each of the administrations treated in this chapter faced significant challenges in their military operations that arose, at least in part, from the quality of the civil-military relationship. Civil-military relations in the administrations discussed in this chapter stemmed in large part from poor analysis and decision-making on the part of the president and his key advisors. George H. W. Bush's administration, according to the conventional wisdom, should have enjoyed the "dream team" of civil-military leadership, with General Powell as chairman and General Schwarzkopf as CENTCOM commander, and both subordinate to Secretary of Defense Cheney, who reported to the president. Yet the decisions the team made mishandled the cease-fire at the end of the First Gulf War and created the conditions leading to Operation Iraqi Freedom more than a decade later.

Clinton created a shambles in Somalia in large part by his personal disengagement from the crisis. Bosnia and Kosovo may have produced the best results possible under the circumstances, but also produced hangovers: the NATO Kosovo Force (KFOR) commitment that endures to this day in Kosovo, and a frozen conflict in Bosnia, where the territory of the former Bosnia-Herzegovina has been fractured into ethnically dominated pieces, including a Serb-majority rump state, Republika Srpska.

The George W. Bush administration's poor civil-military relations contributed to the faulty assumptions adopted to guide the administration's plans for both Afghanistan and Iraq. Poor civil-military relations also led to the adoption of a strategy that was unsupportable given the resources available. The president and his top civilian advisors prized loyalty over competence, and in the process they replaced military leaders such as army chief of staff General Erik Shinseki, whose advice was inconsistent with their own preferences, with yes men who would not provide inconvenient contradictory information.

President Obama walked away from Iraq, preferring disengagement from that blighted country over any positive outcome that might have still been within the United States' grasp. The administration appears headed in the same direction in Afghanistan, although it has slowed the rate of troop withdrawals there in the face of recent requests from the Afghan government. Domestic political advice also seems to trump military advice about national security matters in the Yemen crisis.

The key members of each of these administrations seem to have succumbed to a certain mental habit, described by Daniel Kahneman in his book, *Thinking, Fast and Slow.*[143] As his publisher summarizes,

> Kahneman explains the two systems that drive the way we think. System 1 is fast, intuitive, and emotional; System 2 is slower, more deliberative, and more logical. Kahneman exposes the capabilities—and also the faults and biases—of fast thinking, and reveals the pervasive influence of intuitive impressions on our thoughts and behavior. Then he reveals where we can and cannot trust our intuitions and how we can tap into the benefits of slow thinking.[144]

One of the important insights from Kahneman's work is the revelation that even experts, often steeped in the data and details of their research, when asked for recommendations, will rely on their mind's System 1 to formulate a response, and in the process, ignore the richer and potentially more valuable insights that System 2 thinking might shed on the question.[145] My reading of his research suggests that if future administrations want to minimize the chances of civil-military conflict in planning and executing military operations, they should develop a process that requires all the parties to participate in structured deliberative and decision-making activities: activities in which System 2 thinking can thrive. Such a recommendation may seem far-fetched, but the DoD has a long history of creating processes and programs to maintain critical capabilities. The Cold War–era continuity of government program and nuclear weapons employment procedures are useful examples; in both instances, the DoD created expert staffs to develop the necessary capabilities, exercise them regularly, and assist senior officials in their use and implementation in an actual crisis. The exact civil-military relations program structure and process would be developed as each new president is inaugurated to ensure compatibility with his or her leadership and decision-making style.

Of course, a president might elect not to follow the processes developed to support him/her, and as the commander-in-chief, can do as he/she pleases. Nevertheless, doing the planning and training to support the president with a healthy civil-military relationship may still be useful insofar as it helps the military design and execute operations consistent with the president's intent.

Applying Kahneman's Insights to Four Presidents' Civil-Military Decision-making

The George H. W. Bush administration would have benefited from a process that helped it deliberately evaluate its progress toward its established war goals beyond the liberation of Kuwait (or alert it to the fact that it was straying from its original goals): its progress toward locating, seizing, and securing Saddam's WMD; the percentage of Republican Guard forces destroyed; and the regime's residual capabilities to terrorize and control its population. With sound estimates of progress toward these goals, the administration could have more accurately determined when it was appropriate to announce an end to operations. The same information could have helped the Bush team formulate instructions to General Schwarzkopf for the cease-fire talks (and perhaps sent a different emissary) that would have put Saddam in a position where his regime would have been more vulnerable to future U.S. coercion and control.

The Clinton administration would have been better off with a process that required the president's personal involvement in critical NSC meetings and decisions. Clinton would also have benefited from a process that required his entire national security team—civilian and military—to focus on U.S. interests related to new crises, and the military actions necessary to advance and protect those interests. Such focus would have prevented using military forces in attempts at signaling, as some civilian advisors advocated, and may have enabled the administration to distance itself from changes in UN policy that changed the operating environment for deployed U.S. forces in places like Somalia. Finally, the Clinton White House needed a process—perhaps nothing more than two phone calls from the secretaries of state and defense, respectively, to their subordinates, Ambassador Holbrook and Admiral Smith—that could have helped it compel civil-military cooperation at theater level in places like the Balkans, where theater-level planning and operations were marred by the deliberate isolation of diplomatic and military efforts from each other.

A useful process for the George W. Bush administration would have been one that helped the leadership realize the strategy-resource mismatches they were embracing: a huge nation-building effort in Afghanistan and post-combat stability operations in Iraq (so-called Phase IV operations). President Bush, though nicknamed "the decider," clearly struggled with some decisions including resolving disputes between the state and defense departments on the scope of post-combat activities. He would have benefited from a decision-support process to help him. Such a thing would have been invaluable, especially if it helped the president appreciate the urgency of finding an appropriate solution, potentially saving both Iraqis and the United States the agony that ensued as successive Coalition Provisional Authority chiefs Jay Garner and then Paul Bremer struggled with remits that did not suit the actual conditions on the ground in Iraq.

The Obama administration could have benefited from a process that helped the president and his staff reconcile military-technical analysis (e.g., additional force requirements) provided by the senior uniformed leadership with the domestic political concerns that drove most of the criticism emanating from the NSS and the vice president.

Each of the administrations dealt with in this chapter had a highly organized national security bureaucracy at its disposal, with an established set of policy bodies including interagency policy committees, deputies committees, and principals committees. Despite this apparatus, these administrations experienced important failings in their ability to prosecute military operations.

All the administrations treated in this chapter would have benefited from nimble, System 2–based deliberate (routine) and crisis planning and analysis that could help the president and his senior-most advisors conceive and then test their response plans when new crises emerge. The Defense Department has large analytical organizations at its disposal (e.g., Cost Assessment and Program Evaluation [CAPE]), as do the individual military services. Part of their regular duties should be the development of norms—for example, force-to-population and force-to-space ratios—so that future planners have an empirical basis for sizing their force packages that would help them explain their recommendations to the president and secretary of defense. These organizations should also develop modeling and simulation tools that can be employed quickly to explore different scenarios and military courses of action in response to a crisis. The DoD has many modeling and simulation tools in its inventory, but none that can represent land, air, and naval activities together, and that can be quickly tailored to the enemy, weather, and terrain in an unfolding crisis.

The secretary of defense should never be disappointed by the contingency plans on the shelf, as Rumsfeld indicates he was. To prevent such dissatisfaction, the president, the secretary of defense, combatant commanders, and the service chiefs should agree on a list of such plans, develop them, and review them frequently. The service chiefs' involvement is important because it is their responsibility, under Title 10, U.S. Code, to raise, train, and equip their forces. Understanding what their forces might be asked to do would help them raise, train, and equip the forces appropriately. The combatant commanders are the officers who will employ the forces within their areas of responsibility (AORs). They should be (and have been) involved in the contingency planning process so they have an early sense of what forces will be available to them to address a brewing crisis. The president and secretary of defense should be involved in the effort so that they understand the art of the possible: the various forms of military response that might be mounted, and the time and resources required to do so.

Most important in both deliberate and crisis planning, all the parties should be clear on the chain of causality or theory of victory. There must be a clear description of how military actions will create circumstances that lead to the outcomes the president seeks. The chapter on campaign planning will have more to say on this subject.

Finally, there is the matter of implementation. How would such a process be superimposed upon civil-military relations—especially one that might make military-technical advice more difficult for future administrations to ignore or downplay in favor of other considerations? Congress has an important role in this regard. Congress should be prepared to pass the equivalent of Goldwater-Nichols to impose some order and procedure on the military side of civil-military relations. Congress stepped into the military's domain with Goldwater-Nichols after a series of incidents including the bungled Iranian hostage rescue convinced Capitol Hill that the military could not, on its own, resolve serious issues of interservice cooperation and coordination. The cases presented in this chapter demonstrate that presidential administrations since the Cold War have been unable to make civil-military relations reliably serve the national security needs of the United States. Congress should pass the equivalent of the War Powers Resolution[146] to require the president to work within the deliberative framework intended to support effective civil-military relations by providing structure to advice, collaboration, and decision-making with respect to national security matters. Future presidents are likely to resist, as past presidents have resisted compliance with the War Powers Resolution. Nevertheless, there may be incentives for future presidents to use the civil-military relations framework, especially if future military operations produce failure or mixed results; at least the president can demonstrate he/she used all the tools at the administration's disposal.

Notes

1. Samuel Huntington, *The Soldier and the State: The Theory and Politics of Civil-Military Relations*, 3rd ed. (Cambridge, MA: Belknap Press, Harvard University Press, 1998), 1–2.
2. For a concise summary of the various treatments of civil-military relations analysis found in the literature, see Dale R. Herspring, *The Pentagon and the Presidency: Civil-Military Relations from FDR to George W. Bush* (Lawrence: University Press of Kansas, 2005), 3–22.
3. Commander Dayne E. Nix, "American Civil-Military Relations: Samuel Huntington and the Political Dimensions of Military Professionalism," *Naval War College Review* 65, no. 2 (Spring 2012): 88–105.
4. Richard H. Kohn, "Out of Control: the Crisis in Civil-Military Relations," *National Interest* (Spring 1993/1994).
5. Ibid.
6. The summaries presented in this section of the chapter are drawn from Herspring, *The Pentagon and the Presidency.*
7. Ibid., 298–99.
8. Ibid., 302, 303.
9. *New York Times*, "U.S. Military Chief Is Replaced In the Central American Region," http://www.nytimes.com/1989/07/21/world/us-military-chief-is-replaced-in-the-central-american-region.html.
10. Herspring, *The Pentagon and the Presidency*, 305.
11. *Operation Just Cause: The Incursion into Panama*, U.S. Army Center for Military History, CMH Pub. No. 70-85-1.
12. Herspring, *The Pentagon and the Presidency*, 308.
13. The original plan was codenamed Blue Spoon. It underwent major modifications as OPLAN 90-1 and OPLAN 90-2. See *Operation Just Cause.*
14. Herspring, *The Pentagon and the Presidency*, 308, 309.
15. Wil Grant, "Has Panama Weaned Itself Off Drugs and Cleaned Up?" August 3, 2014, http://www.bbc.com/news/world-latin-america-27945558.
16. *Operation Just Cause*, 48.

17. Herspring, *The Pentagon and the Presidency*, 310, 312.

18. National Security Directive 54, http://nsarchive.gwu.edu/NSAEBB/NSAEBB80/.

19. George H. W. Bush and Brent Scowcroft, *A World Transformed* (New York: Vintage, 1999), 489.

20. Herspring, *The Pentagon and the Presidency*, 315.

21. U.S. Army Center for Military History, *War in the Persian Gulf: Operations Desert Shield and Desert Storm August 1990–March 1991*, CMH Pub. 70-117-1 (2010), 3.

22. Herspring, *The Pentagon and the Presidency*, 314.

23. Ibid., 317.

24. General Schwarzkopf characterizes the October 6 version of the ground offensive plan in almost exactly the same words. He complained to General Carl Vuono, the army chief of staff who was visiting, about having to send a briefing team back to Washington "to brief a plan I don't believe in." See H. Norman Schwarzkopf and Peter Petre, *It Doesn't Take a Hero* (New York: Linda Grey Bantam Books, 1992), 356, 358.

25. Herspring, *The Pentagon and the Presidency*, 319–21.

26. Ibid., 322.

27. Schwarzkopf and Petre, *It Doesn't Take a Hero*, 359–61.

28. Herspring, *The Pentagon and the Presidency*, 323.

29. Daniel Bolger, *Why We Lost: A General's Inside Account of the Iraq and Afghan Wars* (Boston: Eamond Dolan Book/Houghton Mifflin Harcourt, 2014), xxxvi.

30. Schwarzkopf and Petre, *It Doesn't Take a Hero*, 479. According to his account, Chairman Powell asked Schwarzkopf to draft "a set of military conditions that Iraq would have to meet" in order for the cease-fire to hold.

31. Eliot A. Cohen, *Supreme Command: Soldiers, Statesmen and Leadership in Wartime* (New York: Free Press, 2002), 198. Cohen goes so far as to claim the civilians "abdicated" their responsibilities to the military.

32. Herspring, *The Pentagon and the Presidency*, 335.

33. Ibid., 333.

34. Ibid., 336.

35. U.S. Army Center for Military History, *The United States Army in Somalia 1992–1994*, CMH Pub. 70-81-1, 6.

36. Herspring, *The Pentagon and the Presidency*, 343.

37. LTG Montgomery interview, http://www.pbs.org/wgbh/pages/frontline/shows/ambush/interviews/montgomery.html.

38. Herspring, *The Pentagon and the Presidency*, 345.

39. Ibid., 346.

40. Ibid.

41. Ibid., 348.

42. Ibid., 349.

43. David Milne, *Worldmaking: The Art and Science of American Diplomacy* (New York: Farrar, Straus and Giroux, 2015), 425.

44. Presidential Decision Directive 25, http://fas.org/irp/offdocs/pdd25.htm.

45. Herspring, *The Pentagon and the Presidency*, 350.

46. See Walter E. Kretchik, Robert F. Baumann, and John T. Fishel, *A Concise History of the U.S. Army in Operation Uphold Democracy* (U.S. Army Command

and General Staff College Press, Fort Leavenworth, Kansas, 1998). In addition to the details of the crisis that prompted the Clinton administration to intervene, the report provides a deep background on U.S. military involvement on the island.

47. Department of State Office of the Historian, https://history.state.gov/milestones/1914-1920/haiti.

48. Department of State Office of the Historian, https://history.state.gov/milestones/1993-2000/haiti.

49. Herspring, *The Pentagon and the Presidency*, 352.

50. Michael R. Gordon, "Failure of Haiti Operation Backs Initial Pentagon Skepticism," *New York Times*, October 15, 1993, http://www.nytimes.com/1993/10/15/world/failure-of-haiti-operation-backs-initial-pentagon-skepticism.html.

51. Department of State Office of the Historian, https://history.state.gov/milestones/1993-2000/haiti.

52. Herspring, *The Pentagon and the Presidency*, 354.

53. For example, the Vance-Owen plan sponsored by European Community representative David Owen and UN representative Cyrus Vance, and the so-called Contact Group comprised of the United States, Russia, United Kingdom, France, Germany, and Italy.

54. Herspring, *The Pentagon and the Presidency*, 355–57.

55. Ivo Daalder argues the Clinton crisis management policy toward the Balkans had been rendered no longer credible on Capitol Hill, and the increase in Serb offensive operations forced a change in administration policy. See Ivo Daalder, "Decision to Intervene: How the War in Bosnia Ended," Brookings Institution, http://www.brookings.edu/research/articles/1998/12/balkans-daalder.

56. The UNPROFOR-detailed Dutch battalion was intimidated into inaction by Serb forces and failed to protect Bosnian Muslims who had gathered in what was supposed to be a UN-protected safe area. See *The Guardian* account: http://www.theguardian.com/world/srebrenica-massacre.

57. Herspring, *The Pentagon and the Presidency*, 358.

58. PBS *Frontline* story at http://www.pbs.org/wgbh/pages/frontline/shows/military/guys/sbio.html.

59. Herspring, *The Pentagon and the Presidency*, 359.

60. Dayton Accords, Annex 1-A, Article 1, http://peacemaker.un.org/sites/peacemaker.un.org/files/BA_951121_DaytonAgreement.pdf.

61. Richard Holbrooke, *To End a War* (New York: Modern Library, 1999), 327–29.

62. According to Herspring, *The Pentagon and the Presidency* (362), Clark's nomination for SACEUR was not supported by the army; Richard Holbrooke, Sandy Berger, and Strobe Talbott lobbied General Shalikashvili, the JCS chairman, for it.

63. Herspring, *The Pentagon and the Presidency*, 360, 361.

64. Ibid., 363.

65. Ibid., 365.

66. See Anna Di Lellio, "Legendary Commander," www.annadilellio.com/wp-content/uploads/2013/.../Adem-Jashari-NN.pd.

67. Herspring, *The Pentagon and the Presidency*, 365.

68. Ibid., 366.

69. See http://balkanwitness.glypx.com/racak.htm.

70. Text of the resulting Interim Agreement available at http://smsjm.vse.cz/wp-content/uploads/2008/10/sp30.pdf.

71. Herspring, *The Pentagon and the Presidency*, 367.

72. In this they were not successful. The U.S. air fleet flew more than twenty-nine thousand sorties compared to 2,414 sorties for France, 636 for Germany, 1,081 for Italy, and 1,950 sorties for the United Kingdom. See John E. Peters, et al., *European Contributions to Operation Allied Force: Implications for Transatlantic Cooperation* (Santa Monica, CA: RAND Corporation, 2001).

73. Herspring, *The Pentagon and the Presidency*, 371, 373.

74. Ibid., 374.

75. General Hugh Shelton, *Without Hesitation: The Odyssey of an American Warrior* (New York: St Martin's Press, 2010), 399.

76. Lyle J. Goldstein, "General John Shalikashvili and the Civil-Military Relations of Peacekeeping," *Armed Forces and Soc*iety 26, no. 3 (Spring 2000): 387–411.

77. Shelton, *Without Hesitation*, 401.

78. Herspring, *The Pentagon and the Presidency*, 378, 379.

79. Ibid., 380.

80. Shelton, *Without Hesitation*, 402.

81. Milne, *Worldmaking*, 418.

82. Donald Rumsfeld, *Known and Unknown: A Memoir* (New York: Sentinel, 2011), 358.

83. Herspring, *The Pentagon and the Presidency*, 382. In his memoirs, however, Rumsfeld writes (368) that he and Shelton's successor, General Richard Myers, "averaged three or four hours a day together while we held our posts."

84. Ibid., 383.

85. Ibid., 384.

86. Rumsfeld, *Known and Unknown*, 369.

87. Ibid., 359.

88. Ibid., 370–72.

89. Ibid., 373.

90. Condoleezza Rice, *No Higher Honor: A Memoir of My Years in Washington* (New York: Crown, 2011), 314, 316.

91. George W. Bush, *Decision Points* (New York: Crown, 2010), 194.

92. Ibid., 197.

93. Ibid., 205.

94. Douglas J. Feith, *War and Decision: Inside the Pentagon at the Dawn of the War on Terrorism* (New York: Harper, 2008), 134.

95. Herspring, *The Pentagon and the Presidency*, 388.

96. Feith, *War and Decision*, 137.

97. Ibid., 139.

98. Ibid., 140.

99. Rumsfeld, *Known and Unknown*, 683. The secretary of defense also claims a more modest goal, writing, "Our more modest goal was to rid Afghanistan of al Qaeda and replace their Taliban hosts with a government that would not harbor terrorists."

100. Feith, *War and Decision*, 188–94.

101. Ibid., 218.

102. Herspring, *The Pentagon and the Presidency*, 398.

103. Rumsfeld, *Known and Unknown*, 480, 481.

104. Feith, *War and Decision*, 289.

105. Chairman Myers instruction to General Franks, December 19, 2002, in Feith, *War and Decision*, 292.

106. Feith, *War and Decision*, 293.

107. Ibid.

108. Herspring, *The Pentagon and the Presidency*, 399.

109. Joint Operational Planning and Executing System. Joint Pub. 1-02.

110. Rumsfeld, *Known and Unknown*, 439.

111. Ibid., 438.

112. Feith, *War and Decision*, 223.

113. David Knowles, "Rumsfeld: George W. Bush Was Wrong About Iraq," *Bloomberg Politics*, June 8, 2015, http://www.bloomberg.com/politics/articles/2015-06-08/donald-rumsfeld-george-w-bush-was-wrong-about-iraq.

114. The White House, National Security Strategy of the United States, May 2010, preface.

115. Ibid.

116. Ibid., 43.

117. Rick Brennan interview with the author, March 17, 2015.

118. Rick Brennan, "Withdrawal Symptoms: The Bungling of the Iraq Exit," *Foreign Affairs* (November/December 2014): 25–36.

119. Ibid., 29.

120. Ibid., 30.

121. Ibid., 33.

122. Ibid., 34.

123. Ibid.

124. Angel Rabasa, John E. Peters, et al., *Ungoverned Territories: Understanding and Reducing Terrorism Risks* (Santa Monica, CA: RAND Corporation, 2007).

125. Robert M. Gates, *Duty: Memoirs of a Secretary at War* (New York: Alfred A. Knopf, 2014), 338.

126. Ibid., 350.

127. Ibid., 362.

128. Ibid., 474.

129. Ibid., 476.

130. Ibid., 367.

131. Ibid., 368.

132. Ibid., 369.

133. Catherine Dale, "War in Afghanistan: Strategy, Operations, and Issues for Congress," Congressional Research Service, R40156, March 2011.

134. Gates, *Duty*, 490.

135. Ibid., 555.

136. "By with and through" is military shorthand for the various doctrinal relations advisors can create with supported foreign military forces. See Department of the Army Field Manual 3-07.1, Security Force Assistance, May 1, 2009.

137. Adam Taylor, "Four Months Ago, Obama Called Yemen's War on Terror a Success. Now the Yemeni Government May Fall," *Washington Post*, January 20, 2015, www.washingtonpost.com/blogs/worldviews/wp/2015/01/20.

138. CENTCOM Commander's Posture Statement before the House Appropriations Committee, March 5, 2015, www.centcom.mil/en/about-centcom-en/commanders-posture-statement-en.

139. See www.telegraph.co.uk/news/worldnews/middleeast/yemen/11487729/US-special-forces-evacuate-from-yemen-as-security-situation-deteriorates.html.

140. Mark Perry, "U.S. Generals Think Saudi Strikes in Yemen a Bad Idea," April 17, 2015, www.america.aljazeera.com/articles/2015/4/17/US-generals-think-saudi-strikes-in-yemen-a-bad-idea.html.

141. Ibid.

142. CENTCOM Posture Statement.

143. Daniel Kahneman, *Thinking, Fast and Slow* (New York: Farrar, Straus and Giroux, 2011).

144. Publisher's summary available at https://smpl.bibliocommons.com/item/show/268429066_thinking,_fast_and_slow.

145. Kahneman, *Thinking, Fast and Slow*, chapter 22.

146. The War Powers Resolution (also known as the War Powers Resolution of 1973) (50 U.S.C. 1541–1548) is a federal law intended to check the president's power to commit the United States to an armed conflict without the consent of the U.S. Congress.

3

Campaign Planning

campaign—A series of related major operations aimed at achieving strategic and operational objectives within a given time and space. See also **campaign plan**. (JP 5-0)

campaign plan—A joint operation plan for a series of related major operations aimed at achieving strategic or operational objectives within a given time and space. See also **campaign; campaign planning**. (JP 5-0)

campaign planning—The process whereby combatant commanders and subordinate joint force commanders translate national or theater strategy into operational concepts through the development of an operation plan for a campaign. Campaign planning may begin during contingency planning when the actual threat, national guidance, and available resources become evident, but is normally not completed until after the president or secretary of defense selects the course of action during crisis action planning. Campaign planning is conducted when contemplated military operations exceed the scope of a single major joint operation. See also **campaign**; **campaign plan**. (JP 5-0)

As circumstances in Afghanistan, Iraq, Yemen, and elsewhere in the Middle East attest, the United States has lost its ability to design and execute successful military campaigns. They seem to enjoy initial success, but eventually bog down into protracted operations laden with unintended consequences. This chapter describes the key events and influences that, in effect, atrophied the United States' ability to plan and make war and calls for congressional action to restore this vital capability.

The ability to plan a complete, coherent military campaign is among the essential skills for senior military commanders and their civilian masters. Campaigns can be offensive or defensive. Campaigns must be complete in the sense that they deliver the military outcomes the president seeks (e.g., destruction of an enemy regime, abandonment of policies damaging or threatening to the United States, or removal of enemy offensive capabilities such as nuclear, chemical, or biological weapons). Campaigns must also be coherent; the actions planned and their sequence of occurrence should lead logically to the president's desired outcomes, and in the process, not create any new threats to the United States and its interests. The various U.S. military war colleges invest substantial amounts of time teaching strategy and campaign planning to their students. For example, the U.S. Army War College devotes about one-third of its course work to strategy and campaign planning.[1]

The Traditional Ideal: Schlieffen, Soviet Counter-Offensive, U.S. Pacific Campaign

Since the beginning of the era of modern warfare with the First World War,[2] campaign planning evolved as a method to identify and treat specific military problems. The German Schlieffen Plan is a useful example. With it, the chief of the German general staff, Count Alfred von Schlieffen, sought to solve Germany's key military problem: it was situated between two allied states, France and Russia, and if war came, it would have to fight on two fronts simultaneously.[3] The resulting plan that sought to overcome these circumstances was premised on Schlieffen's observations of Russian behavior and troop movements during the Russo-Japanese War, which suggested Russia would be relatively slow to mobilize its forces. Russian deliberate mobilization therefore presented an opportunity to concentrate German military action against France before Russia was prepared to fight. Schlieffen envisioned an envelopment sweeping into France through the Benelux countries in the north to surround Paris and compel France to surrender quickly.[4] Once France collapsed, Germany could turn its attentions eastward, toward Russia.

The chain of causality presented in the Schlieffen Plan was straightforward. German forces would surprise France by attacking through neutral territory, where France's frontiers were least well defended, either destroying French forces in their way or compelling them to retreat. As the German assault continued at a relatively quick tempo, French forces would be driven from their defensive positions and would have to fall back, seeking new defendable ground upon which to reestablish their lines. German forces would continue their turning motion, flanking the remaining French defenders, ultimately

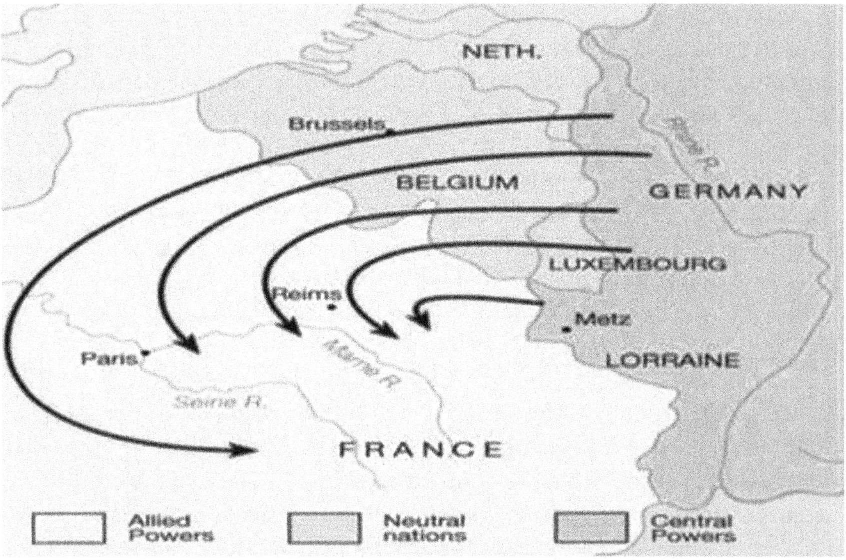

The Schlieffen Plan

encircling Paris. At that point, France could ask for terms or risk destruction of its capital and the remainder of its army. The German theory of victory expected that, with Paris and its army surrounded and threatened with destruction, the French government would prefer to surrender on German terms.

As Germany contemplated the use of military force prior to the First World War, Helmuth von Moltke, Schlieffen's successor, modified the plan in terms of the forces committed to its execution. The rationale and chain of causality within the plan remained the same, however. The German offensive found that Belgian resistance was greater than German commanders had anticipated, slowing the German attack's rate of advance, and buying time for France to recover from the initial onslaught. The German attack was stopped at the Marne River, and Germany's best chance for swift victory vanished.

Soviet Counter-Offensive, Second World War

The Soviet counter-offensive campaign against Nazi Germany during the Second World War had to address two major military problems. The first was liberating Soviet territory from Wehrmacht control. Three German army groups sat astride Soviet territory and had threatened Moscow and Leningrad in the winter of 1942.[5] The second problem was destroying Hitler's regime and prevent Germany from ever again threatening Soviet security.

The counter-offensive campaign contained four main elements. First, a limited defensive counter-attack relieved the pressure on Moscow and Leningrad. Next, the Soviets moved to the offensive, with the destruction of Germany's Sixth Army at Stalingrad during 1942 and 1943, followed by the battle of Kursk the summer of 1943, which destroyed a significant amount of Nazi offensive capability in the process. With the Wehrmacht on the defensive, Soviet forces advanced steadily so that by December, Soviet forces had liberated Kiev. The final phase of the campaign was the invasion of Polish and then German territory, the systematic destruction of German forces, the Nazi regime, and the equally systematic pillage, rape, and plunder of the German populace.

The chain of causality in the Soviet campaign began while Stalin was still on the defensive, rebuilding his military capabilities in factories east of the Ural Mountains. Once new units became available, they would first be employed to drive German units from the doorsteps of Moscow and Leningrad and buy the government some margin of safety. As more military capability became available, it would be massed to destroy the German army groups occupying Soviet territory and deprive Berlin of its offensive capabilities. As German forces were driven westward toward their own frontier, the Red Army would keep up the pressure and, upon entering German territory (East Prussia), begin inflicting enormous damage, driving people from their homes, burning, murdering, and raping—generally terrorizing the fleeing population so that, as it fled west, it would tell its stories of deprivation to everyone they encountered, spreading the terror further. Hordes of civilians on the highways would interfere with German forces' attempts to mount a deliberate defense. Eventually, the continuous pressure of the Soviet advance would fracture German resistance; units would lose their coherence and become vulnerable to battles of encirclement and annihilation, to which many succumbed. The remaining German forces would not be able to defend their capital or their leaders. The Red Army would overwhelm any remaining resistance and assert control over Germans within its occupation sector, as agreed with the other allies, and continue to administer its sector for the foreseeable future.

Actual events proved more difficult. German resistance was fierce, and the Red Army faced some significant setbacks. Nevertheless, the logic of the campaign prevailed. The Red Army succeeded in destroying three Wehrmacht army groups, threw the remaining German units onto the defensive, eventually overwhelmed them and seized Berlin from its defenders. Soviet occupation was shaped by the London Protocol of 1944, which established the occupation zones and practices for the British, American, and Soviet forces. In accordance with the 1945 Potsdam Conference agreements, the Soviet Union created the Soviet Military Administration for Germany, which ruled from the German capitulation in May 1945 until the founding of the German Democratic Republic in 1949. The Soviet Military Administration provided virtually all of domestic governance functions initially, and it gradually ceded powers to the Germans, creating German administrative and governance capabilities consistent with Soviet preferences. And the Soviets were good at it. Historian Nicholas Stargardt notes,

> Berliners were surprised by the speed with which the Soviet military authorities restored the food supply and set about clearing the streets, repairing the tram and the underground lines and restarting the gas, electricity, and water supplies.[6]

The Group of Soviet Forces in Germany, a traditional, combined arms force, reinforced the authority of the Soviet Military Administration.

U.S. Pacific Campaign in the Second World War

The United States campaign in the Pacific during the Second World War had to address two military problems. The first was breaking the brutal imperial Japanese domination over the peoples within the Japanese "co-prosperity sphere"—that part of the western Pacific under Tokyo's military, political, and economic control. Japan's armed forces had invaded China in 1931 and subsequently extended their influence into other Southeast Asian territories. Japanese forces were well trained and equipped, sustained in large part by the Imperial Navy. The second military problem was destroying the militarist regime responsible for the war crimes and misery that had ensued. The campaign plan had at least four main elements:[7] first, defeat the Japanese navy so it could no longer attack U.S. forces or sustain Japanese forces deployed throughout the region; second, seize islands suitable for supporting air bases within range of the Japanese islands, and attack the home islands with sustained bombing; third, liberate allies from enemy occupation (e.g., Philippines); and fourth, assault the home islands and defeat the defending forces.

The chain of causality underpinning the campaign plan began by defeating the Japanese navy. Once the Japanese navy was no longer a factor, U.S. and allied forces would enjoy greater freedom of action. Japanese navy forces would no longer threaten U.S. and allied forces or oppose their landings on Japanese-occupied territory. Nor would the Japanese navy be able to support the imperial Japanese army and its multitude of garrisons across the many islands in the region. Garrisons not defending islands suitable for airfields could be left to languish until later. Next, the United States would land its own forces to seize and clear islands in range of Japan and build air bases on them. These bases would be home to U.S. Army Air Force units conducting a long-duration bombing operation against Japanese industries and the civilian population, which had been dubbed an enemy population because of its support of the war effort. As the bombing took its toll on Japan's capability to wage war, and its unsupported garrisons marooned on islands of little military significance began succumbing to disease and hunger, U.S. and allied forces could begin liberating territory from Japanese control. British, French, Dutch, and other colonial possessions, as well as the Philippines, saw their Japanese occupiers taken prisoners, and, in some instances, the commanders of those forces tried for war crimes. Once there was no longer any threat of deployed Japanese forces interfering with a U.S. invasion of the home islands, such an invasion became the priority. Additional forces, victorious in Europe, would be shifted to the Pacific to take part in the effort.

The campaign unfolded more or less according to plan. The battles of Midway and Guadalcanal early in the war did major damage to the Japanese navy, depriving it of critical aircraft carriers, pilots, and troops. The Japanese grip on occupied territories was broken through assault landings and brutal combat. Japan itself was subjected to sustained bombardment, which delivered the destruction anticipated by the campaign plan.[8] The striking departure from the original plan was, of course, the atomic bomb attacks on Hiroshima and Nagasaki. Despite claims of unconditional surrender, strategic surrender talks took place,[9] a bloody invasion was not required, the emperor retained his throne, and General Douglas MacArthur oversaw the U.S. occupation of Japan from the Dai-ichi Palace.

According to the Department of State, the occupation, which ended in 1952, contained three distinct phases. The first, from 1945 to 1947, was punitive in nature. It saw war crimes trials, the disestablishment of the Japanese armed forces, prohibition of former military officers to enter politics, and land reforms. In 1947, the Supreme Command Allied Powers dictated a new constitution which, among other things, limited future Japanese armed forces to purely defensive capabilities and delegated greater authority to parliament. Later phases of the occupation dealt with economic and material shortages.[10]

The U.S. occupation forces were primarily combat formations of the army, Marines, and navy, with newly created special staff sections. U.S. combat units were initially deployed as area commands, and there were many transfers of authority from units that were returning to the United States to newly arrived formations.[11] The special staff sections were organized as economic and scientific, civil information and education, natural resources, public health and welfare, government, legal, civil communications, statistical and reports, and civil intelligence. These staff sections established relationships with their Japanese counterparts, working to reform their authorities and activities for an eventual handoff of responsibilities, and to address the needs of the Japanese people as the country began its recovery from the destruction wrought by the war. The British and Commonwealth allies also provided occupation units and administered their own areas in similar fashion.

American Expeditionary Force in World War I

American involvement in the First World War presented a somewhat different focus from the traditional campaign plans described above. The European powers had been fighting since 1914, and the inconclusive slaughter left the armies of the western front bogged down along a more or less static set of trench lines. In the east, Czarist Russia had collapsed in the midst of revolution, and the Germans had imposed a tough peace upon the Bolshevik regime that had seized power from the more moderate Kerensky government several months after the United States' declaration of war.[12] The military problem confronting General John J. Pershing, the commander of the American Expeditionary Force (AEF), was resisting the British and French demand that U.S. soldiers serve as replacements to fill the decimated ranks of British and French units and find a way to make militarily significant contributions that would make the United States influential at the peace settlement following the Central Powers' defeat.[13]

To address the first part of the problem, Pershing simply refused "amalgamation," as placing U.S. soldiers in allied units was known. Both President Woodrow Wilson and Secretary of War Newton Baker supported his decision. However, when pressed, Pershing did assign the 93rd division, which was manned by African Americans, to the French for deployment.

For the second part of the problem, General Pershing found locations along the front where U.S. units could be committed to good effect. Paris was still within range of a German offensive, so preventing breakthroughs (and closing them quickly if they occurred) was imperative. Moreover, counterattacking such breakthroughs could destroy significant numbers of German units and force the survivors to abandon the offensive.

The chain of causality for the AEF campaign had three major military links: first, get the AEF to Europe quickly to bolster British and French morale, and keep their armies in the fight (there had been mutinies in both French and British units in response to ghastly casualties); second, inflict major losses on German forces and remove any last hopes that Germany could win; and third, make the allies beholden to the United States in order to ensure U.S. influence in the subsequent peace talks. There was also a political link in the chain of causality, based on President Wilson's Fourteen Points, which, among other things, provided for a European territorial settlement and principles for the conduct of open post-war international relations that did not depend upon the size and might of individual states.[14]

Both Hindenburg and Ludendorff, the German commanders, believed they still enjoyed the means to continue operations despite the losses their forces had sustained and planned a major offensive to try to destroy the allied armies and take Paris before U.S. forces arrived in significant numbers. Ludendorff had assembled 3.5 million men by March 1918.[15] The Lys offensive, begun in April, found the British in serious trouble as the Germans exploited the collapse of a Portuguese division, maneuvered eight divisions of their own to positions of advantage against the British, and looked like they might push the British forces back to the sea. The German advance lost momentum, however, as hungry soldiers stopped fighting and foraged for food in the countryside. The German army nevertheless continued to pursue the offensive, launching major operations throughout the spring and summer, culminating in the July offensive.

U.S. forces were instrumental in halting these German offensives and in the allied counter-offensives that followed, reducing potentially dangerous salients in allied lines and inflicting significant losses on the enemy. Chateau Thierry, St. Mihiel, Meuse-Argonne, Belleau Wood, and the Marne were all successful major U.S. engagements. As summer faded, the German army still stood well forward on French territory, but its troops were exhausted and many malnourished; discipline in the ranks was frayed as well.

In a September 27 public address in which Wilson elaborated on the Fourteen Points, and according to H. W. Brands, Wilson "essentially asked the German people to overthrow their government," the president said:

> We are all agreed that there can be no peace obtained by any kind of bargain or compromise with the governments of the Central Powers, because we have dealt with them already and have seen them deal with other governments that were parties to this struggle. . . . They have convinced us that they are without honor and do not intend justice. They observe no covenants, accept no principle but force and their own interest. We cannot "come to terms" with them. They have made it impossible. The German people must by this time be fully aware that we cannot accept the word of those who forced this war upon us.[16]

Wilson in effect attacked Clausewitz's "iron triangle" of the sovereign, the army, and the people.[17] German society was already rife with unrest, roiled by communism, social democracy, and lesser political forces. It was also experiencing the privations of extended warfare and the British blockade. An invitation from the president of the United States to overthrow the German government simply sought to amplify the effects of forces already at work in Germany. Wilhelm II, the sovereign, abdicated, seeking asylum in the Netherlands. He and his government had in any case been marginalized by Hindenburg and Ludendorff, who had come to orchestrate all aspects of the German Empire's activities in support of the war. The German army, of course, was marooned in France.

By September 28, Ludendorff concluded the war was lost and advised Berlin to seek terms. Discussions began on October 4. By October 12, the Germans agreed to the Fourteen Points as the basis for an armistice.[18] The British and French, however, balked at the deal, thinking it too generous and especially skeptical of the notion of open negotiations. Wilson threatened a separate, German-American peace if the allies did not agree. The threat carried the day with the allies, and a final cease-fire took place on November 11, 1918.

The AEF campaign during the First World War stands out from the other examples explored in this chapter as much more political-military in nature. That is, the abdication of the kaiser, collapse of his government, and loss of popular support were at least as important as the exhaustion of the German army in achieving allied victory—a critical point Hindenburg missed in describing the war's end in his memoirs.[19] The chain of causality included important military links, such as effective operations that destroyed significant numbers of enemy units. But the chain also included political links that contributed to the kaiser's decision to abdicate, and that may have accelerated the political crisis that swept through German society.

As one can see, the campaigns examined briefly in the preceding pages shared several attributes. First, they each sought to treat a specific type of military problem: how to avoid simultaneous war on two fronts, how to eject an invading army and destroy the regime that sent it, how to eject an occupying army and destroy the regime that sent it in a theater dominated by water and large distances, and how to use military accomplishments to influence post-war outcomes. Each campaign plan also attempted to apply force in a systematic, ruthless way to achieve its goals. All that changed with the advent of the Cold War.

Cold War Influences on Campaign Planning

The United States' monopoly on nuclear weapons disappeared August 29, 1949, with the Soviet detonation of its first nuclear device. The People's Republic of China followed suit on October 16, 1964. The presence of nuclear weapons made the Cold War different from any previous political-military struggle. First, it offered the prospect of an adversary inflicting grave damage—perhaps even total annihilation—on the United States, even as the

United States was on the threshold of victory against its enemies. Second, because of the vast destructive potential of long-range missiles topped with nuclear warheads, campaigns like those of the world wars were widely believed to be impossibilities. If the Warsaw Pact attempted to overrun Western Europe, Washington might destroy scores of Soviet cities while at the same time decimating the invading armies with tactical nuclear weapons—while sustaining similar devastation at home and on the battlefield. For the U.S. government at the time, the perceived threat in any crisis that it might escalate to the use of nuclear weapons imposed constraints on political objectives and on military actions that might be mobilized to help attain them.

Coincident with the beginning of the Cold War, many European states began divesting themselves of their colonies. Often, the newly independent states faced bitter internal conflicts over who would rule and what form governance would take. Because the colonial masters were typically democratically governed countries, democracy attracted limited numbers of adherents, and Marxism-Leninism was often popular because it seemed to offer solutions for post-colonial societies beset by many problems. Sometimes the product of these post-independence political disputes was civil war. Indochina proved a novel and tragic exception. Rather than granting full independence to the states within the French Union, France sought to reassert its control, and the First Indochina War was the result.

It did not take long for these post-colonial conflicts to become proxy wars between the United States and its allies, on one hand, and the Soviet Union and its client states, on the other. These proxy wars found the United States and the West supporting political factions that claimed to be democrats, and the Soviets supporting Marxist-Leninist factions pursuing "wars of national liberation." In cases where democrats were in short supply, the United States often supported autocrats as long as they professed to be anti-communist.

In some cases, the proxy wars featured communist and putatively democratic states competing against each other: North Korea against South Korea, and North Vietnam versus South Vietnam. In other cases, the local government confronted an internal threat from communist insurgents. Some communist theoreticians conceived of all of these various efforts as systematic steps in a process they called "peoples' war."[20] The first step in peoples' war was to cultivate popular support, typically through a propaganda effort. As support grew, the movement would raise funds through soliciting donations, robbery, and extortion, building a financial base that would support buying arms and equipment. Next, the movement would begin low-level attacks on government facilities to steal supplies and equipment, and to demonstrate the government's vulnerability. Gradually, as the group's initial successes attracted more volunteers, it would also launch a campaign of assassination against local officials and popular authority figures (e.g., schoolteachers, civic leaders). Eventually through intimidation of the populace and successful operations against the government, the movement would be able to establish a liberated zone from which to stage its future actions. As the government weakened and its military became demoralized, the revolutionary movement would become correspondingly more capable, and transition from guerrilla, hit-and-run actions to conventional units able to meet and destroy government military units in decisive battle.

The communist revolutionary warfare approach benefited from the U.S. fear of taking too much direct military action against a Soviet proxy state. A lesson the United States learned from the Korean War was that successful operations in the territory of North Korea prompted Chinese intervention when UN forces advanced too close to the Chinese border. In the case of the Vietnam War, Washington was prepared to bomb Hanoi and mine its harbor at Haiphong but gave no serious consideration to an invasion to destroy the North Vietnamese army and then seize the capital. Large-scale search-and-destroy operations in South Vietnam targeting Viet Cong and North Vietnamese army units were one thing; Washington believed decisive military action against the North Vietnamese regime was another. This reluctance reflected Washington's fear of provoking Soviet military action, a fear that was well grounded. In September 1968, General Secretary Leonid Brezhnev announced the Brezhnev Doctrine:

> When forces that are hostile to socialism try to turn the development of some socialist country towards capitalism, it becomes not only a problem of the country concerned, but a common problem and concern of all socialist countries.[21]

Although the doctrine was announced explicitly to deter reform or moderation within the communist regimes of the Warsaw Pact members, it may also have been applicable if the United States and South Vietnam had launched a regime-threatening campaign across the demilitarized zone. At minimum, the Soviet Union would have increased its military assistance to North Vietnam. If Moscow concluded that a poor showing in defense of North Vietnam in some way endangered its grip on the Warsaw Pact, it might have taken more aggressive steps to defend its client state.

The resulting approach Washington took to these proxy wars was to shore up the local government, build its ability to defend itself, and try to improve its legitimacy in the eyes of its own population. Typically, U.S. efforts focused on counter-insurgency. In the process, traditional campaign planning was supplanted by the orthodoxy of counter-insurgency thinking. The specific, detailed chain of causality that lies at the heart of traditional campaign planning was replaced by the one-size-fits-all notion from counter-insurgency theory of winning the trust and support of the population. The next chapter will treat this subject in greater detail, but it is useful here to know that counter-insurgency generally involved separating the populace from the insurgents so the people couldn't be terrorized or otherwise influenced by them, helping the government carry out reforms that restored its legitimacy and redressed popular grievances, and training and equipping police and security forces so they were capable of defeating the insurgency. In most cases, counter-insurgency during the proxy wars also involved U.S. forces in direct combat with the enemy, often to buy time for the supported regime to train and deploy capable forces of its own. Military planning in the context of counter-insurgency focused on the tasks just described. Success was creating a local government that was effective at national defense, internal safety and security, and at meeting the needs of its people: a tall order, as events from the Cold War to today (Vietnam to Afghanistan) have illustrated.

There were also a number of crisis interventions over the course of the Cold War, but they did not rise to the level of campaigns; rather, they were individual operations. Most of them transpired within the context of the West's Cold War strategy of containment. The United States landed Marines in Lebanon in 1958 at the invitation of the government to address insurrection aided by outside support. In 1964, the United States sent transport aircraft to deliver Belgian paratroopers during the Congo crisis following independence. In 1965, President Johnson sent Marines and paratroopers to the Dominican Republic to prevent the rebellion there from producing a communist government like that in Cuba. In 1981, major elements of El Salvador's army were transported to U.S. training sites for retraining to enable them to pacify the insurgency that was ravaging the country. In October 1983, President Reagan sent Marines and army airborne forces to Grenada to restore public order at the request of the Organization of East Caribbean States. In 1989, President Bush sent forces to help the Philippines defeat a coup attempt.[22]

Only two operations took place outside the context of containment of Soviet influence. The first of these was Operation Eagle Claw, the failed Carter administration effort to rescue American hostages held in Iran. According to researchers Stuart Koehle and Stephen Glick, the military planning for the operation was complex enough to invite confusion and failure, and the units and equipment chosen to execute the operation were sub-optimal for the tasks and conditions involved.[23] Indeed, Operation Eagle Claw was one of the episodes of poor military performance that motivated Congress to pass Goldwater-Nichols six years later. The second operation was Operation Just Cause, which removed Manuel Noriega from power in Panama.

One can see from the descriptions above how both the proxy wars of the Cold War era and most of the crisis response operations undertaken by U.S.

forces during that era required no comprehensive campaign planning in large part because they did not attempt to treat the major military problem driving these events, Soviet international activities and influence. That job was left to the strategy of containment. Surely, though, the standoff between the North Atlantic Treaty Organization (NATO) and the Warsaw Pact must have produced a major campaign planning effort in NATO headquarters. But it did not. The next section explains.

NATO and the Warsaw Pact

From the very beginning (1949), NATO members had to balance two critical considerations. The first was economic development and recovery from the devastation left by the Second World War. The second was national security and collective defense. The Marshall Plan, the United States Productivity Mission, and the Ford Foundation helped with West European recovery and development.[24] NATO played a major role in national security and collective defense by organizing the initial twelve members' efforts into a coherent strategy.

The principal military concern was Soviet aggression and the spread of communism. Recent events at the time had not been promising. In Poland, communists seized power in 1947. Czechoslovakia suffered a communist-led coup in February 1948. Moscow created the Warsaw Pact in May 1955 and intervened in Hungary to crush its anti-communist revolution in November 1956. Witnessing such events, Soviet aggression was no abstraction for the defense ministers of Western Europe.

As the alliance matured and grew, it adopted a policy of deterrence and dialogue.[25] Under this policy, the allies sought political détente with the Soviets while providing an adequate defense for themselves. And the alliance's military posture was decidedly defensive. The 1950 mid-term planning document's call for ninety divisions was considered unaffordable two years later at the 1952 North Atlantic Conference in Lisbon, Portugal. The allies agreed instead to develop nuclear weapons for its forces at the 1955 conference,[26] substituting them for manpower and reinforcing the alliance's deterrent posture. There were few strategic reserve ground forces, however. The United States usually maintained two corps in Germany and was committed to provide a total of ten divisions, much of which were stationed in the United States, and their personnel had to be flown to Europe in event of a crisis. For most of the other members, once they committed the forces within their individual corps boundaries, there were no more.[27]

The ground forces of the alliance varied in quality but shared a common limitation. Those along the inter-German boundary, the center stage for any future confrontation, were organized into corps comprised of two or three divisions. These units generally had good tactical mobility that would enable them to maneuver against an adversary. However, they lacked operational and strategic mobility because they were all tethered to the warehouses, pipelines, and resupply points that provided their logistics and combat service support. This was not a system designed to attack toward Moscow; instead, it presented a deliberate defense of allied territory. The specific details of the defense evolved through the years. "Follow on forces attack" (FOFA), a concept adopted in the mid-1980s by the allies, sought to halt an enemy assault and also attack the enemy's second echelon ground forces from the air before they could reinforce the Warsaw Pact units in contact with NATO and threaten a breakthrough of allied defenses.[28] FOFA remained largely conceptual, however, and the alliance failed to acquire all of the military hardware necessary to realize it. Moreover, FOFA was really theater doctrine; its concepts were never developed at NATO Army Group level in enough detail (e.g., reflecting the specific, local chain of causality, terrain, and the capabilities and limitations of the specific allied units that would do the fighting) to represent campaign plans.[29] Thus, politicians and military commanders alike were in a position where they simply hoped that NATO's defensive posture would dissuade the Warsaw Pact from an attack and that, if an attack came, the steady attrition of Warsaw Pact forces and the threat from the allies to use nuclear weapons would cause Moscow to halt military operations.

These expectations were undermined, however, by the fact that some of the allied forces held relatively light reserves of ammunition, fuel, and other expendables critical to warfare. The prevailing thinking at the time was that, if war came, the conventional battle would be short, and the alliance would resort to tactical nuclear weapons to stop enemy penetrations of allied territory. There was no need, in such circumstances, for large war reserve stockpiles.[30] These circumstances meant NATO, especially in the central region along the inter-German boundary, was betting everything in the event of an attack on its ability to use its nuclear weapons to stave off defeat. Few senior military officers and fewer allied politicians had a clear sense of how the Atlantic Alliance would shape events once nuclear weapons began detonating. The uncertainties associated with going nuclear were so great that few officers could imagine—though U.S. doctrine attempted to[31]—just what would be required to fight on contaminated ground in order to exploit the effects of a nuclear strike on the enemy.

The alliance's contingency plans and exercises themselves were tactical (i.e., corps units and below) and defensive in nature. Virtually all of them were responses to anticipated aggression from the Soviet Union and its Warsaw Pact accomplices. Military staffs planned deliberate defenses fought for the most part from prepared defensive positions and defensive counter-attacks to throw back aggressors that managed a breakthrough. Planning focused on maintaining the integrity of the alliance defenses, rather than on a campaign to halt the enemy assault and defeat Warsaw Pact forces in detail. The final step in NATO's defenses was the use of nuclear weapons. NATO's arsenal extended from short-range battlefield nuclear artillery to cruise and ballistic missiles capable of reaching targets deep in Soviet territory. Theories abounded on the deterrent functions of these weapons and the likely behavior of the antagonists in the event of a nuclear exchange. NATO campaign planning—such as it was—basically stopped with battlefield use of such weapons. Most officers simply hoped limited use of these weapons would lead promptly to a negotiated settlement, although catastrophic outcomes were certainly possible.

The Cold War endured for decades, arguably from Winston Churchill's Fulton, Missouri, "Iron Curtain" speech on March 5, 1946, until the breaching of the Berlin Wall in November 1989. Campaign planning went dormant as a military skill during those forty-three years. Several generations of military officers served through this period, in circumstances in which they did not do comprehensive campaign planning. What military actions they carried out took place as simpler operations. Although some of these operations (e.g., Eagle Claw) were complex, they did not match the scale of military problems that confronted Schlieffen, Moltke, Pershing, or the campaign planners of the Second World War. Along the way, campaign planning became a forgotten art. Efforts to revitalize it have not been entirely successful.

Jedi Knights and Strategists

The military services historically have operated mid-level schools that prepared career officers for future command and staff responsibilities. These officers, typically majors and lieutenant commanders, spent six months at the Armed Forces Staff College or a year at the service equivalent institutions. The curriculum varied somewhat from school-to-school, but generally sought to make its graduates tactically proficient and capable of commanding battalions/squadrons, or serving on higher-level service or joint staffs. This focus meant that officers frequently learned the prevailing practices and procedures rather than deepening their own capabilities for military analysis and problem solving. In the early 1980s, the army moved to create a follow-on course to supplement its Command and General Staff School. The School of Advanced Military Studies (SAMS) was the result, and it began operations in 1983. The school's task was to teach operational art, which in prevailing doctrine was the splice between tactics and strategy. The other services followed suit thereafter. SAMS graduates were sometimes dubbed "Jedi Knights" after the Star Wars movie characters. Alumni were instrumental in planning Operation Desert Storm and later operations.[32]

Above the staff colleges stand the war colleges: the service war colleges, National War College, and Industrial College of the Armed Forces. Only the most successful officers attend, and these colleges are clearly the gateway to general officer stars. Their purpose is to prepare officers for national-level service and high-level command. Students are supposed to learn grand strategy and political-military affairs. In practice, however, some of the war colleges have become rewards for successful battalion command. Officers enjoyed a leisurely schedule with ample time to reconnect with their families after the demanding, time-consuming responsibilities of command. Officers could graduate without mastering the substance of the curriculum. Representative Isaac Newton (Ike) Skelton, long-time chairman of the House Armed Services Committee (HASC), was among the most critical members of Congress and one of the stoutest advocates for professional military educational reform. He sought to restore academic rigor, grades, and uniformity of education for senior officials throughout the executive branch of government.[33] He died before such reforms could be accomplished fully.

Most recently, the army has expanded its FA-59 strategist career field in an attempt to revitalize its capability to develop strategy.[34]

All of these efforts—adding additional schools, reforming professional military education, and expansion of a potentially important career field—are likely to fail if they do not somehow get back to the basics of campaigning and campaign planning while also analyzing the problems that have bedeviled U.S. forces during Phase IV operations. That is, the military must not only relearn how to take the guidance from the president and secretary of defense, identify the key military problems that must be solved in responding to that guidance, establish the chain of causality between the use of force and the resulting outcomes, and then allocate the available resources and apply them in a way to bring about the desired outcomes. They must also treat the specific issues that arise in Phase IV, conducting stability operations, quelling insurgencies and insurrections, and restoring public order. These are not easy tasks. They have confounded dozens of the U.S. military's ablest generals and admirals.

Gray Zone Conflicts and All That

To be sure, it is more difficult to identify the military problem to be solved in the gray zone conflicts described by Nora Bensahel and LTG David Barno

(U.S. Army, Ret.) in their blog.[35] Afghanistan, with its Taliban government hosting an alien terrorist organization, al Qaeda, which had just carried out horrendous attacks on the United States, is a case in point. As General Stanley McChrystal described during an interview with the *Wall Street Journal*, the United States rushed in to Afghanistan:

> The first thing is we didn't do due diligence before we went in. We didn't understand the problem to the depth that we needed to. We didn't take the time to do it, and we didn't nurture the experts.[36]

Even if the United States had not rushed into Afghanistan as General McChrystal observed, the military problems would be nested in a complex of non-military issues common to failed states that had to be addressed in order to defend and advance American interests. Syria's civil war is a telling example. The regime is engaged in a civil war against multiple Syrian factions, that seek to unseat President Assad. There are strong sectarian overtones involved, with the Alawite regime facing threats from Sunni and Shia factions, including the Sunni extremist group ISIS. The competition between and among these groups presents a tangled blur of competing goals and plenty of deceit. The warring factions have also attracted international support, including support from Iran, Russia, and China for the beleaguered regime, and Turkish and other international support for some of the militias. For a U.S. president regarding the resulting sea of suffering and dangerous instability, what are the administration's objectives? Beyond minimizing human suffering, what do U.S. interests compel an administration to try to do? If the answer is "support peace and regional stability," what is the military problem to be solved? Must U.S. forces train, equip, and advise some of the combatants, and, if so, which ones, and to what end? Or, if one imagines U.S. interests require American armed forces to defeat some enemy, which ones, and what defines success and why?

Analysis of gray zone conflicts often results in calls for "whole of government" solutions, calling on other, non–Department of Defense (DoD) departments within the executive branch of government to deploy personnel and resources to dangerous regions of the world. The State Department and the U.S. Agency for International Development are among the most frequently cited examples. The problem confronting all of the non-DoD departments is that they cannot compel their personnel to accept dangerous overseas assignments, instead relying on volunteers. Moreover, as Condoleezza Rice noted in her memoir, the supply of foreign service officers is relatively small: around eleven thousand.[37] Nor are contractors a viable solution. Although they have at times outnumbered U.S. military personnel in Iraq,[38] they are not subject to the Uniform Code of Military Justice and thus can be compelled to undertake only those services specified in their contracts. And most contractors provide sanitation, facilities, and their maintenance and operations (e.g., dining facilities, shower facilities, temporary quarters) or security services. They are an unlikely source for military governance and similar whole of government services.

For military campaign planners, the emphasis on whole of government solutions can lead (and has led) to wishful thinking. Just as General Tommy Franks and his CENTCOM planners ignored post-conflict planning in Iraq, imagining some other entity would handle the details, expectations of substantial contributions from non-DoD departments can cause problems in gray zone conflict analysis, with military planners assuming—mistakenly—that they can avoid unpleasant non-military tasks by leaving them for some other agency to pick up. Key tasks left unassigned, like the post-combat phases of Operation Just Cause and Operation Iraqi Freedom, are potential prescriptions for enduring instability and prolonged, futile U.S. military engagement.

So, what might a modern, ideal type campaign look like? Consider the following example.

Imagining a Future Campaign Plan: A Russian Invasion of the Baltics

This scenario presents a possible future crisis and the military requirements to counter it, including NATO's physical and psychological preparedness to respond appropriately, and the outlines of an allied campaign plan to do so.

Moscow alleges that one of the Baltic States is mistreating the members of its Russian minority and vows to come to their assistance. Russian news media present the Russian public with a torrent of contrived stories about the mistreatment and miserable plight of their relations at the hands of the Baltic country's majority. Soon thereafter, Russian military forces begin conducting exercises short distances from the border. At the same time, the Baltic country begins experiencing telecommunications and electricity failures. Senior NATO officials conclude Russian President Vladimir Putin is attempting to recover territory lost at the dissolution of the Soviet Union.

In the early hours of the morning, Russian forces seize a border town and an area of about fifty square kilometers along the frontier to create a safe zone. Moscow then broadcasts an invitation for ethnic Russians to come to the safe zone for protection against their Baltic oppressors.

The invasion triggers the NATO Article Five security guarantee. The alliance thus faces consequential choices: come to the aid of its member state and risk war with Russia, or fail to honor the Article Five promise and do severe damage to the credibility of the Atlantic Treaty and its value as a deterrent against aggression.

The military problems the campaign must treat are as follows:

- Save NATO and restore its deterrent qualities, especially in Russian eyes.
- Restore the territorial integrity of its Baltic member.
- Use military force with great care to limit the scope of the war with Russia.
- Impose costs on Moscow high enough that Putin will be discouraged from similar actions in the future.

Saving NATO and restoring its deterrent qualities would require at least two things not clearly present among the alliance membership today. The first is awakening members of the alliance to the threat of aggression Moscow poses and improving allied military capabilities in response—and, importantly, building consensus for action if an attack appears likely. Second, the alliance must reposition some of its forces along the eastern edge of allied territory to deter Russia from attempting limited objective hybrid attacks to seize elements of allied territory like those it undertook in Ukraine. Rapid reaction forces held some distance from the NATO-Russian frontier are not enough. They would create conditions where Russia could seize territory and present the alliance with a fait accompli, providing the most cautious allies with incentives to dither and delay consensus for action. The consequences from such delays could undermine the alliance itself.

Restoring the territorial integrity of the Baltic partner is a second military imperative. No solution to the crisis that did not restore an alliance member's territory to it would be satisfactory, and would provide stark evidence that the alliance did not do everything in its power to satisfy the Article Five security guarantee. Restoration of the Baltic state's territory involves counter-attacking and ejecting Russian forces from the land they had seized. Such an operation would require a mix of military and police forces. The military would lead the counter-attack to force Russian elements to withdraw. The police would follow, restoring civil order, and looking for Russian stay-behind elements embedded within the Russian minority.

Using military force carefully is imperative to managing risks of escalation. The allies must calibrate the use of force to limit the likelihood of provoking a catastrophic expansion of violence from Moscow. Elements of this effort include communicating to Moscow that defense of its territory is a core interest to NATO, but also that allied forces have no other objectives beyond the restoration of the pre-war status quo, and then demonstrating the allied commitment to its cause by mounting a prompt and vigorous military response on a scale that overwhelms the Russian forces on the scene.

Imposing costs on Moscow for its aggression is important to shaping Russia's post-conflict deterrence calculus. Significant damage and losses are essential to cause Moscow to conclude that the costs of the effort outweighed the potential benefits that might have accrued from it, had it succeeded, and that future efforts are simply too risky, given Western resolve and military prowess. Getting Russian leaders to reach such conclusions requires the alliance to punish Moscow by destroying high-value military assets, but not so many so as to provoke Moscow to widen the war or intensify its offensive into allied territory.

Both imposing costs and using force carefully require a "fight and communicate" approach to addressing Russian aggression.

Chain of Causality/Theory of Victory

Vladimir Putin has concluded that the United States and the European members of NATO have lost their nerve and will suffer any humiliation to avoid direct hostilities with Moscow. Therefore, a vigorous allied military response would enjoy the advantage of surprise. Any successful military action would satisfy the expectations of Article Five.

In order to be both limited in scope and sufficient to cause President Putin to withdraw his forces, military force should be focused on an asset that the Kremlin prizes highly: Kaliningrad, with its two military airfields, Russia's only warm water port beyond the Black Sea, and its Baltic Fleet headquarters.

If NATO inflicts significant damage on Kaliningrad military facilities and forces, and communicates that it is determined to sweep Russian forces from Baltic territory, Russia will withdraw its forces and negotiate a settlement.

The campaign unfolds with simultaneous attacks on Kaliningrad military airfields, garrisons, harbor facilities, and ships of war. NATO air forces destroy Russian air forces on the ground; combinations of cruise and ballistic missiles destroy fixed targets and war ships.

Meanwhile, allied ground forces swarm into the Baltic state and support local forces in finding, killing, or capturing Russian agents and military personnel in Baltic territory. This process includes screening the local ethnically Russian minority for agents who might be hiding within the group.

A potentially decisive aspect of the NATO approach is its immediate resort to a military response without an initial attempt to negotiate with Moscow. This departure from Vladimir Putin's likely expectations demonstrates to the Kremlin that any invasion of allied territory is different from all other security threats to the alliance and that the commitment among members to defend allied territory is unanimous.

Execution of the Campaign

Retaliation for Russia's attack begins immediately, as the following paragraphs describe. Concurrently, NATO mounts a global media blitz, including the Russian population, to tell the truth about the situation in the Baltics, broadcasting interviews with Russian minority members, and showcasing their homes and the quality of their lives. Frontline news video lets global viewers see the battlefield.

SACEUR provides overall military direction for the campaign. United States Air Force Europe (USAFE) serves as the NATO combined force air component commander, integrating allied air force capabilities to prosecute the attacks on the Kaliningrad targets.

United States Army Europe (USAREUR) serves as the NATO combined force land component commander, integrating allied ground forces for operations to eject Russian forces from the Baltic state.

United States Navy Europe (NAVEUR) coordinates maritime operations in the Baltic, Black Sea, Mediterranean, and Atlantic. The principal focus of naval efforts is to contain the Russian navy in the Baltic and Black Seas and defend against wider naval operations in the Atlantic and Mediterranean.

Putin's immediate reaction to the Kaliningrad strikes is predictable; he orders an increase in the speed of the invasion into the Baltic state and denounces NATO for "its unwarranted and cold-blooded act of aggression." The allies stand their ground, and after ninety-six hours of pummeling from the air, the battle damage assessments from his generals in Kaliningrad convince Putin he must stop the destruction of some of his most valuable military capabilities. He communicates his desire for talks to the allies.

The North Atlantic Council sends its reply: The strikes will continue until all Russian forces depart the Baltic state.

Despite its initial success, the Russian invasion has begun to lose its momentum as NATO forces counter-attack. Russian forces in the Baltic state find their perimeter shrinking. The commander expresses his concerns of being cut off from his supporting units across the border in Russia. Putin orders a withdrawal. Allied intelligence learns of the order and uses its surveillance capabilities to confirm Russian forces are crossing back into Russian territory.

The alliance signals that it is prepared to discuss events with Moscow. Finland offers to host the talks in Helsinki. Putin demands that the talks take place under the auspices of the United Nations. The NATO secretary general, president of the United States, and other allied leaders insist on Helsinki. Putin concedes, and a cease-fire prevails.

At the Helsinki talks, the allies insist that Russia pay for the damage it caused during its operations in the Baltic state. The allies also require a promise that Russia will renounce military action against NATO. Putin demands security guarantees and reimbursement for the losses sustained in Kaliningrad. An eventual compromise results.

Because the crisis was short-lived and the Baltic state in question did not suffer any significant degradation of its civil administration, Phase IV operations are short and limited in scope. Allied armed forces make a sweep of the state's territory to confirm that all Russian forces have indeed withdrawn. The Baltic state's police and internal security forces conduct a similar sweep of the population to discover any stay-behind agents and others whose separatist sentiments may incline them to acts of terrorism. After deliberations with the Baltic state's government, NATO's North Atlantic Council endorses a recommendation from the Military Committee to leave a residual military presence in the country to deter future aggression from Moscow. The remaining NATO forces return to their home garrisons.

Conclusions and Observations

It is understandable, given the circumstances of the Cold War, how traditional campaign planning expertise fell by the wayside. The Cold War standoff, with its prominent nuclear weapons dimension, created uncertainties to which U.S. officials responded with caution and a set of constraints on military action and thus campaign planning. The wars of national liberation that the West confronted saw the substitution of counter-insurgency theory for chain of causality analysis and campaign planning. The irregular warfare and gray zone conflicts that have come to typify many of today's wars illustrate why the traditional approach to campaign analysis has become more complicated; the military problem requiring treatment may be obscured by an opaque, dense tapestry of many non-military factors fueling a conflict. In addition, chapter 2 described how the quality of civil-military relations can influence an administration's ability to respond to a crisis, and how the way the administration formulates the overall challenge can have a significant effect on military planners' ability to identify key military problems that must be treated in order to deliver the outcomes the president desires.

The military's own efforts to reinvigorate strategy and, with it, the ability to plan and execute successful campaigns are laudable but unlikely to succeed without congressional help. The military has had multiple opportunities since September 11, 2001, to sharpen its campaign planning skills and has delivered disappointing outcomes at best. Moreover, the military's record of internally driven reform is a poor one.[39] Congress, however, intervened effectively in 1986 with the Goldwater-Nichols DoD Reorganization Act, which was intended

> to reorganize the Department of Defense and strengthen civilian authority in the Department of Defense, to improve the military advice provided to the President, the National Security Council, and the Secretary of Defense, to place clear responsibility on the commanders of the unified and specified combatant

> commands for the accomplishment of missions assigned to those commands and ensure that the authority of those commanders is fully commensurate with that responsibility, to increase attention to the formulation of strategy and to contingency planning, to provide for more efficient use of defense resources, to improve joint officer management policies, otherwise to enhance the effectiveness of military operations and improve the management and administration of the Department of Defense, and for other purposes.[40]

It is time for Congress to consider a follow-up effort in the spirit of Goldwater-Nichols with an emphasis on strategy and campaign planning. Congress led the way before to good effect, clarifying the roles of the military departments, the combatant commanders, the joint staff, and the secretary of defense. If the military is to reestablish its ability to do campaign design and planning appropriate to today's circumstances and conflicts, it will need help from Congress, this time with emphasis on reinvigorating professional military education in campaign planning, especially teaching how to identify the appropriate military problems to be solved in a given emergency and alternative theories of victory/chains of causality for solving them. Congress should also direct the secretary of defense to integrate the resulting improved capability for campaign planning with the efforts to improve civil-military relations and the provision of military advice to the president described in chapter 2.

Perhaps above all else, any effort to revitalize campaign planning should emphasize Phase IV operations. As this chapter has shown, the Soviets were successful in World War II in part because they created institutions to administer their occupation zone and because they had a sound plan for transferring power gradually to a German communist government—all backed up by the coercive power of the Group of Soviet Forces, Germany. The United States was successful in Japan for similar reasons. It had a good plan, implemented in phases, with appropriate expertise beyond regular military functions in terms of justice, governance, economics, and the like. Combat arms occupation forces initially backed up these capabilities. More recently, Phase IV operations have been treated as an afterthought, or dismissed as another institution's responsibility, with tragic results. If the U.S. military is to perfect its campaign planning capabilities, it must devote serious attention to planning Phase IV operations.

As the imagined scenario of future Russian aggression in the Baltics suggests, there is work to be done in the alliance to reach consensus about the threat Moscow poses and an agreed course of action to address it (e.g., the details of the allied campaign plan to deter Moscow and defend alliance territory). The United States should be leading the effort to reach such a consensus. Now that Russia has broken the NATO-Russian Founding Act[41] through its actions in Ukraine and Crimea, NATO has greater latitude to shore up its deterrent posture along the alliance's eastern frontier, and should begin doing so. The president should engage the other allied heads of state on the matter and, in particular, demand that they increase their defense spending to reach the NATO target of 2 percent of gross domestic product. It is time for the European members of the alliance to make deterrence and defense a priority again.

Notes

1. According to the 2013 Army War College Student Handbook (37–38), http://dde.carlisle.army.mil/DDE_documents/StudentHandbook_2013.pdf.

2. This is my own determination. It was not until 1914 that armies had communications, mobility, and firepower reflecting today's capabilities in these domains.

3. This description treats the substance of the plan. Christopher Clark has pointed out that, written in 1905, the plan may have been an argument for more conscription since it employed more forces than Germany had at that time. See Clark, *The Sleepwalkers: How Europe Went to War in 1914* (New York: HarperCollins, 2012), 216.

4. Ian Senior, *Invasion 1914: The Schlieffen Plan to the Battle of the Marne* (New York: Osprey, undated e-book edition), chapter 1.

5. Richard Overy, *Russia's War: A History of the Soviet War Effort, 1941–1945* (New York: Penguin Books, 1998), 99–124.

6. Nicholas Stargardt, *The German War: A Nation Under Arms 1939–1945* (New York: Basic Books, 2015), 542.

7. U.S. Army Center for Military History, http://www.history.army.mil/html/reference/army_flag/ww2_ap.html.

8. United States Strategic Bombing Survey: Japan's Struggle to End the War, 1 July 1946, https://www.trumanlibrary.org/whistlestop/study_collections/bomb/large/documents/pdfs/68.pdf.

9. Paul Kecskemeti, *Strategic Surrender: The Politics of Victory and Defeat* (Santa Monica, CA: RAND Corporation, 1957), chapter 6.

10. Department of State, Office of the Historian, https://history.state.gov/milestones/1945-1952/japan-reconstruction.

11. U.S. Army, *Chronology of the Occupation 15 August 1945–31 March 1946*, http://www.history.army.mil/documents/8-5/8-5.htm.

12. Charles B. Macdonald, *World War I: The U.S. Army Overseas*, Office of the Chief of Military History, U.S. Army, chapter 18, http://www.history.army.mil/books/AMH/AMH-18.htm.

13. Ibid., 383.

14. H. W. Brands, *Woodrow Wilson* (New York: Times Books, Henry Holt, 2003), 89.

15. Macdonald, *World War I*, 385.

16. Brands, *Woodrow Wilson*, 95.

17. Carl von Clausewitz, *On War*, translated and edited by Michael Howard and Peter Paret (Princeton, NJ: Center of International Studies, Princeton University, 1989).
18. Macdonald, *World War I*, 400.
19. Paul von Hindenburg and Frederic Appleby Holt, *Out of My Life*, Volume II (1921) (Kessinger Publishing's Rare Reprints), 270–74.
20. See, for example, Vo Nguyen Giap, *Peoples' War, Peoples' Army* (New York: Praeger, 1962).
21. http://www.britannica.com/EBchecked/topic/79119/Brezhnev-Doctrine.
22. Congressional Research Service, *Instances of Use of United States Armed Forces Abroad, 1798–2014*, Report Number R42738, September 15, 2014.
23. Stuart L. Koehl and Stephen Glick, "Why the Rescue Failed: There Was More to Operation Eagle Claw's Failure in the Desert of Iran than Jimmy Told Us," *American Spectator*, July 1980, http://spectator.org/articles/34807/why-rescue-failed.
24. H. G. Schroeter, *Americanization of the European Economy: A Compact Survey of American Economic Influence in Europe Since the 1800s* (New York: Springer, 2005), chapter 2.
25. NATO, "The Future Tasks of the Alliance" (Harmel Report), December 13, 1967, http://www.nato.int/cps/en/natohq/official_texts_26700.htm.
26. John A. Reed Jr., *Germany and NATO* (Washington, DC: National Defense University Press, 1987), xxi.
27. Theoretically Britain and France were exceptions, with forces outside the alliance defensive structure. After 1966, France's military had no role in the "layer cake" defenses of the central front. As a practical matter, planners at the time doubted a Soviet offensive would allow the time for NATO to deploy additional forces.
28. U.S. Congress, Office of Technology Assessment, *New Technology for NATO: Implementing Follow-on Forces Attack*, June 1987, UNT Digital Library, http://digital.library.unt.edu/ark:/67531/metadc39886/.
29. John E. Peters, "Evaluating FOFA as a Deterrent," *RUSI Journal* (December 1987): 39–46.
30. Author's conversations between 1978 and 1980 with Center Army Group (CENTAG) and III German Corps planners.
31. U.S. Army doctrine had, through several iterations, sought to identify force requirements for the nuclear battlefield. See John L. Romjue, *From Active Defense to AirLand Battle: The Development of Army Doctrine 1973–1982* (United States Army Training and Doctrine Command Historical Office, June 1984).
32. http://usacac.army.mil/sites/default/files/documents/cace/350-1_CGSCCatalog.pdf.
33. https://www.govtrack.us/congress/members/ike_skelton/400377.
34. Lieutenant Colonel Charles Moore, "What's the Matter with Being a Strategist (Now)?" *Parameters* (Winter 2009–2010).
35. See http://warontherocks.com/category/strategic-outpost/.
36. "Leadership Lessons from the Generals," *Wall Street Journal*, December 9, 2014.
37. Condoleezza Rice, *No Higher Honor: A Memoir of My Years in Washington* (New York: Crown, 2011), 316.
38. Congressional Research Service Report R43074, May 17, 2013.

39. See John E. Peters, *The U.S. Military: Ready for the New World Order?* (Westport, CT: Greenwood Press, 1993).

40. Goldwater-Nichols DoD Reorganization Act of 1986, http://history.defense.gov/Portals/70/Documents/dod_reforms/Goldwater-NicholsDoDReordAct1986.pdf.

41. The 1997 Founding Act on Mutual Relations, Cooperation and Security between NATO and the Russian Federation guaranteed the borders of the signatories who, among other things, pledged not to station nuclear weapons or a permanent military presence on the territory of the new states. See https://www.armscontrol.org/act/1997_05/jm.

4

Counter-insurgency and Irregular Warfare

counter-insurgency—Comprehensive civilian and military efforts designed to simultaneously defeat and contain insurgency and address its root causes. Also called COIN. (JP 3-24)

irregular warfare—A violent struggle among state and non-state actors for legitimacy and influence over the relevant population(s). Also called IW. (JP 1)

COUNTER-INSURGENCY (COIN) and irregular warfare constitute opposite sides of the same coin. Because the overwhelming majority of insurgents practice irregular warfare, the COIN forces must be able to defeat it. There is, however, much more to irregular warfare than the Department of Defense (DoD) dictionary of military terms suggests. Irregular warfare is typically a complicated, shadowy affair carried out in the midst of the local population. Indeed, the British general, Sir Rupert Smith, refers to it as "war amongst the people."[1] Chairman Mao Zedong characterized the populace as the sea in which his guerrilla fish swam.[2] In other characterizations, the populace is the prize.[3] The insurgents seek to alienate the people from the government and reorient the population to view the insurgents as the legitimate heirs to power. Irregular forces typically employ kidnapping, murder, and mass casualty events (e.g., truck bombs) against the government and its allies. They embed themselves within the population to make it more difficult for the government's COIN forces to attack them without injuring innocent members of the population around them. When they choose to engage the government and its allied military partners directly, irregular warriors usually conduct hit-and-run attacks and avoid pitched battles, where they are often at a disadvantage.

COIN and irregular warfare have been around since ancient times.[4] The United States arguably has had long experience with both, having employed irregular warfare itself during the Revolutionary War and having attempted COIN in the Philippines episodically from 1901 onward, and irregular warfare from the early nineteenth century in the form of the Barbary pirates and even earlier in conflicts with the American Indian tribes. Why, then, have COIN and irregular warfare become such stumbling blocks for presidential administrations from John F. Kennedy to the present? This chapter examines the influences that have shaped U.S. thinking about COIN and irregular warfare, and offers an explanation and a prescription for future operations. The chapter begins with a summary of the "canonical cases" of COIN: the cases this author and his fellows were steeped in during Reserve Officer Training Corps (ROTC) classes at university in the mid-1960s and subsequently rediscovered by General David Petraeus and his colleagues several decades later. Next, the chapter examines the suitability of these cases for guiding current U.S. military thinking about insurgencies and irregular warfare. Finally, the chapter presents an alternative approach that would prepare U.S. forces for COIN and stability operations (post-conflict operations to provide peace, security, and stability until civil authorities can take over) against adversaries practicing irregular warfare.

Counter-insurgency and Irregular Warfare in Colonial Settings

As noted in chapter 3, the Europeans began experiencing unrest in their colonies in the years prior to the Second World War, and, in part prompted by the United States, began divesting themselves of their empires in the aftermath of the war. Many of the insights concerning COIN resulted from European efforts to preserve order, grant independence, and prevent their former colonies from falling into communist hands.

Britain: Malayan Emergency and Mau Mau Rebellion

Two British COIN campaigns have been especially influential on U.S. military thinking on the subject. The first case involved the British colony of Malaya, officially the Malay Federation, between 1948 and 1960.[5] The colonial administration confronted a communist insurgency growing within the Chinese population of the region: a population comprising 38 percent of the total populace.[6] A twenty-six-year-old named Chin Peng led the insurgency, which he conceived as containing three phases. In phase one, his proven guerrillas, many of whom had fought alongside the British Force 136 (a unit of the British Special Operations Executive conducting unconventional warfare

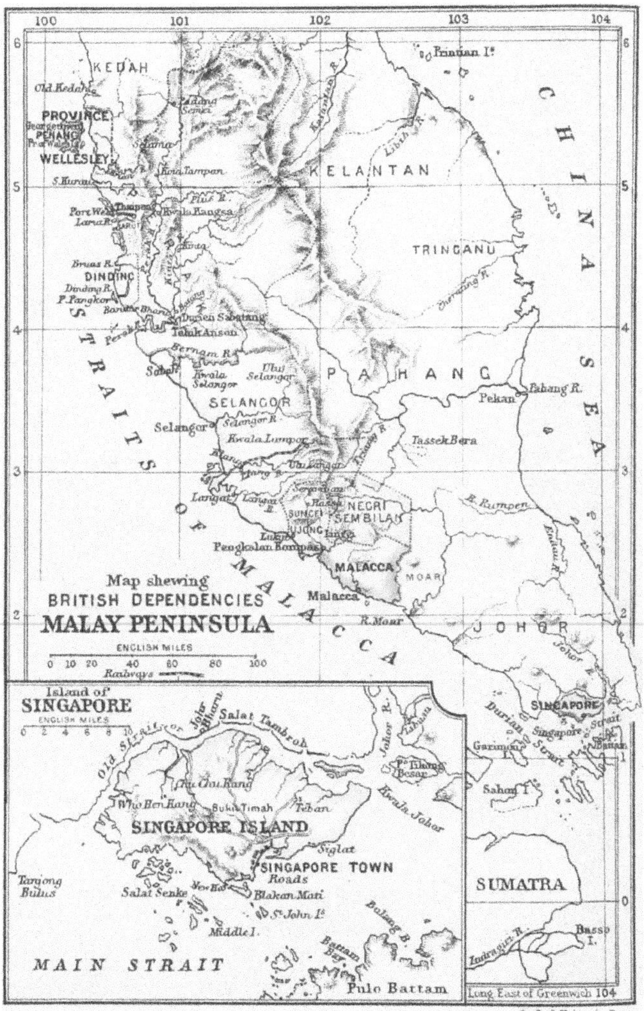

British Dependencies of the Malay Peninsula (1888)

Source: Charles P. Lucas, A Historical Geography of the British Colonies, vol. 1, p. 134 (from Wikimedia Commons)

in Southeast Asia during the Second World War), would attack lonely rubber plantation estates, tin mines, and police and government officials in small towns and villages. These acts, he believed, would force the British to evacuate rural areas and move their populations to towns.[7]

Phase two was to create liberated areas in the spaces evacuated by the British. These liberated zones would be home to guerrilla bases where new members could be recruited and trained for the striking forces. A second,

British Kenya.

Source: National Geographic Society Maps Collection (Public Domain)

larger force—the "masses movement," or Min Yuen—would also be recruited from across all of Malayan society, penetrating all social strata to gather information useful to the striking forces and to support those forces with food, equipment, and arms. The Min Yuen would also serve as a recruitment pool from which future striking force members could be drawn.[8]

Phase three, as Chin Peng envisioned it, would see territorial expansion from the liberated areas with his communist army attacking towns, villages, and railways. The Min Yuen would undermine the economy and security, creating widespread chaos. Finally, the army would take on British forces directly. Chin Peng imagined his army would be backed by Mao's Red Army if necessary.

The Mau Mau rebellion in British Kenya was distinctly different.[9] It lacked the political coherence of the Malayan Emergency and the outside support that Chin Peng and his confederates enjoyed (e.g., periodic advice from Red Chinese army officers). The grievances that mobilized Mau Mau support were multiple: underemployment in Nairobi, discontent channeled into militant politics, and the historical practice of the tribal elites, driving the poor from their lands in the Native Preserves. Moreover, mechanization of the large European farms made the squatter labor force an unwanted burden for the planters.[10] Mau Mau supporters espoused "land and freedom" as their objective, but the Lari Massacres of March 26, 1953, in which Mau Mau killed ninety-seven Home Guards and their family members and wounded another fifty, according to Daniel Branch, were motivated by older grievances arising from 1940 land seizures.[11] The violence gradually drove the Kikuyu, the tribal people at the center of the emergency, into either the Mau Mau or the loyalist camp. Loyalists were generally not supporters of colonial rule but concluded they had a better chance of surviving and protecting their interests by aligning with the colonial authorities than confronting the Mau Mau movement on their own.[12]

In both Malaya and Kenya, the British pursued an approach involving five main efforts: first, to separate the guerrillas from their support bases within society; second, to impose controls to limit the population's ability to provide support to the enemy; third, to offer the populace something more attractive than what the guerrillas were offering; fourth, to create rewards and incentives to support the government and to quit the guerrillas; and fifth, to destroy the enemy fighters.

Separating the Guerrillas from Their Support Bases

In Malaya, the High Commissioner, Sir Henry Gurney, acted vigorously in this regard. He ordered the uprooting and resettlement of six hundred thousand Chinese squatters who lived on the edge of the jungle and were a key source of support (involuntarily and voluntarily) to the enemy. Gurney moved them to "New Villages" where they held title to their land, but in locations remote from the Communist Terrorists, or CTs.[13] Sir Henry used Emergency Regulation 17D, which established authority to uproot entire villages and move them into detention camps if there was evidence of support to CTs.[14] The High Commissioner employed the Banishment Act to deport to China more than ten thousand Chinese in 1949 alone.[15] Gurney also tried to cultivate support among the Chinese by encouraging them to join the Malayan Chinese Association, an organization that worked to improve conditions in Chinese villages. The High Commissioner's effort helped increase membership to one hundred thousand and raised \$2.5 million for village improvements.[16] Colonial authorities used the security forces to protect unpopulated areas, cutting lines of communications between CTs and the villages, and forcing the CTs to fight. Eventually, the authorities also built forts deep in the jungle where most of the aborigine population lived. The forts offered the aborigines security and protection from CT coercion.[17]

In Kenya, the process began with arrests and mass detentions at the onset of the emergency, October 20, 1952. Jomo Kenyatta, leader of the Kenya African Union (KAU), and five of his lieutenants were among those arrested. According to Branch, the Mau Mau movement was decapitated from the beginning. The simultaneous mass detentions and forced removal of population disrupted the movement's coordination and planning, and also prompted thousands to flee to the forest in the highlands of Central Province to avoid capture.[18] The Provincial Administration mobilized the Kikuyu tribal chiefs and their Tribal Police to collect intelligence and to carry out punitive sanctions on Mau Mau adherents and their supporters.[19] Chiefs were given arrest quotas to fulfill, and failure to meet quota could result in a chief losing his job.[20] Some one hundred thousand Kikuyu were removed from the Rift Valley where the large European farms were located and returned to Central Province.[21] The colonial authorities conducted large-scale screening operations, including Operation Anvil in Nairobi in April 1954. Such operations employed violent interrogations of the population and detention of suspected Mau Mau supporters. Sometimes they also involved mass clearance of city areas suspected of general Mau Mau support.[22]

Impose Controls to Limit the Population's Ability to Provide Support

In addition to separating the guerrillas from the population, the authorities in both Malaya and Kenya employed a variety of controls, including abridgement of rights and novel restrictions that left the population with little to offer the enemy in terms of their support. As one Malayan told the authorities, "If the Communists come with a gun and demand food, we are killed if we refuse to help. Do something by law which makes it impossible for us to help them."[23]

In Malaya, colonial authorities authorized detention without trial for up to two years, mandatory death sentences for suspects found carrying weapons, searches without warrants, curfews, restrictions on movement and food controls, and a nation-wide registration program that resulted in every person having an identity card with a photo and thumbprint on it.[24] Such steps reduced the demand side of the equation, the number of CTs available to coerce the populace to provide food, equipment, and information. The colonial authorities also published a list of restricted goods—all things CTs in the jungle would prize—and required transfer and sale of goods on the list to be documented with a receipt that included the identity of the recipient and his address. Then there was Operation Starvation, a novel campaign that severely restricted access to food. In restricted areas, food was rationed at the village level.[25] It could not be taken from the home. Even miners, accustomed to taking their lunches to work with them, had to return home at noon to eat. Restaurants and cafes were closed. Food could not be moved at night (when it was most vulnerable to hijacking). Rice had to be cooked upon receipt and consumed immediately, lest it spoil.[26] These steps collectively reduced the supply side of the equation, the sources of material support available to be exploited by the Communists. In addition, the colonial administration gradually built up its security forces. During the 1952–1954 counter-attack phase of operations, there were forty thousand British and colonial troops deployed, twenty-five thousand Malaya police, and fifty thousand special constables and 350,000 home guardsmen.[27] Special Branch, nominally part of the police, enjoyed "unlimited powers" in prosecuting their internal security operations.[28] Not counting the Home Guard, the security forces were thus able to generate a force-to-population ratio of twenty-three security force members per one thousand of population: an important statistic treated later in this chapter.

Kenyan colonial authorities implemented similar detention, control, and curfew policies to those found in Malaya. They also emphasized locally appropriate penalties for Kikuyu who supported Mau Mau. These included communal fines, seizure and forfeiture of livestock and other possessions, and closing of coffee factories, markets, shops, and schools serving villages and locations suspected of collaboration.[29] The village chiefs and their Tribal Police became the local enforcers for the Provincial Administration, and they were supplemented in July 1952 by the creation of a militia. The militia would evolve into a Home Guard not unlike that in Malaya, but never so large (roughly twenty-six thousand at its height). It protected prominent loyalists, gathered intelligence, denied the enemy access to food and supplies, and policed the Native Reserve.[30] The authorities also built a network of fortified guard posts manned by Home Guard. Initially they often collaborated with the Mau Mau warriors who came to attack them, but eventually the guard posts and the Home Guard who defended them became oases of security and stability, and villagers would often spend the night there.[31] The colonial authorities embraced "screening" or mass interrogations to examine most of the Kikuyu population. During screening, Tribal Police, Home Guard, and police and army operations combined to sort the population into Mau Mau supporters and loyalists, and the authorities created new villages and forcibly relocated the population to them. Loyalists were assigned their spaces within the new village and forced to build their own accommodations. The villages were located near Home Guard posts for security, and most had perimeter fences. Livestock, food stores, and similar goods were carefully controlled.

Mau Mau supporters were moved to punitive villages where they were compelled to build their own, more austere dwellings. In Daniel Branch's words, the Home Guard maintained an "iron grip" over the villages.[32] All of these actions combined to produce a "swing of control" wresting coercive power from Mau Mau and providing sufficient security so that loyalists could safely oppose the Mau Mau movement.[33]

Offer the People Something More Attractive Than the Guerrillas Can Offer

Colonial officials launched multiple efforts to give all the peoples of Malaya a stake in society and the current government. The Chinese squatters resettled in New Villages were given the land and title to it.[34] Among other things, title to land gave them something they could borrow against, improving their economic position in society. Authorities also publicized heavily the British promise of independence[35] and pointed to India and Burma as evidence that the promise was sincere.[36] Government officials were also supportive of the Malayan Chinese Association, which, as noted, bettered the conditions for the Chinese minority and helped to integrate its members into broader Malayan life. One of the more influential actions may have been creating War Councils at the state and district level and inviting Chinese and other minorities to participate in them.[37] In a similar action, the administration created New Village councils to which residents could be elected.[38] The War Council memberships removed opaqueness from the government's management of the emergency and gave residents some influence on that effort. Membership in a New Village council was an exercise in democratic governance. Both initiatives let people glimpse the potential of their society to provide a harmonious, peaceful life in the future. The authorities also embarked on a major public education program and sent six hundred Asians to England for a year-long course to be trained as teachers.[39] Another important initiative was the new citizens code adopted on September 14, 1952. It conferred citizenship on 1.2 million Chinese and 180,000 Indians.[40] General Sir Gerald Templer, the first "Supremo" to command both the military and police forces in the colony, launched a program of three-day war tours that gave ordinary people an idea of how the government was prosecuting the war and produced a sense that government was working. One of the stops along the tour included a visit to the Employee Provident Fund, where the participants could see their pension accounts, indicating to each of them that they had a material stake in the current government.[41] Finally, there was Malayanization, a process that brought more indigenous figures into government, devolved more responsibilities to them, and also produced the first national elections for a Federal Legislative Council.[42]

On the Communist side, during an October 1951 conference, Chin Peng conceded that his plans had failed. His directive, reorienting CT efforts, noted that terrorism had not worked and had alienated the people it was meant to liberate. Vandalism in the mines and plantations threatened people's livelihoods, the directive observed. The directive ordered an end to all events that caused "inconvenience" to the masses.[43] In the end, both the Communists and the colonial administration were offering liberation to the residents of Malaya. The difference in the offers was profound, however. The near-term byproduct of the CT campaign was murder; the byproducts of government efforts were improving security, economic development, and expanded access to education.

In Kenya, the authorities emphasized impending de-colonization.[44] Otherwise, officials concentrated on efforts that would provide loyalists with jobs or improve their economic standing. Jobs became more abundant as security improved.[45] Agricultural reform in the form of the Swynnerton Plan encouraged cash crops and established credit facilities and farm planning services. The plan created a landed elite and an agricultural laboring class.[46] A land consolidation plan complemented agricultural reforms. Land holdings were typically fragmented, making farming inefficient. The consolidation plan, which was placed largely in loyalist hands, first measured one's holdings, then reallocated a single plot of the same size, and issued a land title, which enabled mortgage lending.[47] Loyalists also found the path to permits required for many forms of commerce simplified for them. The Tribal Police and Home Guard both offered jobs. Some resettlement to the more prosperous Rift Valley helped Kikuyu find work. Overall, as Branch notes, "land reform and labor opportunities dovetailed with Kikuyu political thought."[48]

Mau Mau could not compete. As noted earlier in this chapter, the movement's organization and planning had been preempted at the onset of the emergency, and mass detentions of suspected supporters drove many more into the forest to avoid arrest. Mau Mau thus lacked a coherent program to support its "land and freedom" slogan. Moreover, life in the forest in the central highlands was hard; many who fled there succumbed to pneumonia and starvation. By February and March 1954, mass surrenders were in progress.[49] The colonial authority's promise of independence coupled with concrete benefits for loyalists trumped the empty slogan of Mau Mau.

Create Rewards and Incentives to Support the Government and Quit the Guerrillas

Officials in Malaya created a system of rewards and inducements designed to influence both the population at large and the CTs and their supporters. A cash-for-guns program supported a network of double agents who sold out arms caches.[50] The government also offered rewards for information leading to the killing or capture of CTs, and, in a novel twist, CTs themselves were eligible if they sold out their comrades.[51] September 1949 was an important month in terms of rewards and inducements.[52] The government announced its terms of surrender for CTs. It also authorized police rewards for informants. Thereafter, Lam Swee, one of the Communist political commissars and thus important to the insurgency, defected and penned a pamphlet titled "My Accusation," which criticized the insurgency in detail. Thousands of copies were scattered in the jungle to incentivize CT surrender.[53] Rehabilitation camps were created, which offered CTs the prospect of humane treatment when they surrendered and eventual reentry into Malayan society.[54] Next, the government announced larger rewards for the capture of enemy leadership, including their surrender—that is, a senior member of the insurgency could benefit by turning himself in and collecting a reward.[55] Authorities scattered safe conduct passes over areas thought to be home to CT camps. Civilians could earn rewards by assisting in the surrender of CTs.[56] Perhaps one of the biggest inducements to quit the insurgency came from Chin Peng himself. By 1958, fortune had turned against his movement and his forces were badly depleted. Chin Peng, operating from a remote redoubt in Thailand, authorized the remaining CTs to surrender and provided them each with $420 in severance pay.[57]

In Kenya, authorities in 1955 began offering amnesty for crimes committed early during the war to induce Mau Mau fighters to surrender.[58] Otherwise, officials appear to have emphasized penalties supporters faced if they did not turn themselves in. Without a screening certificate, the product of submitting to interrogation, former Mau Mau supporters could not get jobs, secure permits, or acquire a land parcel. Those who failed to surrender faced the prospect of having their land and livestock confiscated.[59] Moreover, the benefits accruing to the loyalists were becoming well known. The weight of the incentives favored surrender. Surrender, however, was not synonymous with defeat. As Branch observed, "Despite their military defeat by both loyalists and the colonial state, Mau Mau sympathizers understood Kenya's impending independence as their ultimate victory over the British and loyalists."[60]

Destroy the Enemy

Despite the emphasis many authors have placed on other facets of COIN campaigns, destroying the enemy remained key to both Malayan and Kenyan colonial officials. On December 12, 1948, security forces surrounded a rubber plantation along the Batang Kali River and killed twenty-four unarmed men there.[61] One month after declaration of the emergency in Malaya, police killed Lau Yew, the "only CT with any pretentions to military tactics," in effect depriving Chin Peng and his insurgency of military advice.[62] As the police, military, and Home Guard grew to meet the challenge, these forces moved from their initial defensive posture to the counter-attack in 1952. The offensive efforts to destroy the enemy crystalized in the form of the Bourne Plan, conceived by Lt. General Sir Geoffrey Bourne. In it, the Special Air Service (SAS) deployed to the mountains to unseat the CTs from their jungle bases and attack them where they lived. Army and police field forces dominated the known CT areas, inflicting losses upon the enemy whenever the opportunity presented itself. Police and civil authorities enforced the restricted items list and the food embargo provisions.[63] Disease and starvation combined with combat losses to decimate the CTs. And decimation understates the level of destruction delivered. Of the approximately twelve thousand members of Chin Peng's forces, 6,698 were killed. Another 2,696 surrendered, and 1,286 were captured. Thus, approximately 89 percent of the CT force was killed, captured, or surrendered. A total of 2,819 were wounded.[64] By 1958, only 250 CTs operated in Malaya.[65]

In Kenya, at the declaration of emergency, "Kenya police and army concentrate[ed] on the destruction of the growing guerrilla army in the forests and the Home Guard was expected to act as the government's eyes and ears in the Native Reserve of Central Province."[66] By March 1954, the Home Guard was 25,600 strong in Central Province, operating a network of fortified guard posts. In addition, Kikuyu Guard Combat Units were formed with capabilities of the regular army units they replaced.[67] The Home Guard accounted for 50 percent of Mau Mau casualties.[68] From mid-1953 onward, Mau Mau forest fighters were forced from their preferred bases near their home areas and driven deeper into the forest. The capture of Kimathi, the last Mau Mau general, on October 21, 1956, left the armed element of the movement leaderless.[69] The majority of twenty-five thousand deaths from 1952 to 1956 were Mau Mau activists. One hundred-seventy African members of the armed forces were killed, along with eighteen hundred African opponents of the insurgents. Thirty-two European settlers and sixty-three European combatants were killed.[70]

France: Indochina and Algeria

The military problems confronting France in Indochina and Algeria were different from those Britain faced in Malaya and Kenya. In Indochina, the first task for France was to reestablish its authority. Its officials and colonial forces had been interned by the Japanese in the last weeks of the Second

World War. The Potsdam Conference at the end of the war restored France's Indochinese colonies, but it would be October 1945 before thirty-five thousand French forces could arrive in Saigon.[71] In the meantime, fourteen hundred French military personnel released from custody by the British, who had been sent to oversee the Japanese surrender south of the sixteenth parallel, ran amok, murdering Vietnamese civilians. By the time France reestablished its military presence throughout the country, the Viet Minh had grown to be a consequential guerrilla force and Ho Chi Minh had declared an independent Democratic Republic of Vietnam (DRV). Following the defeat of Chiang Kai-shek's Chinese Nationalist forces by Mao's Communist army in 1949, the Communist government began shipping U.S. military hardware from the defeated KMT forces to the Viet Minh. General Vo Nguyen Giap, commander of the Viet Minh army, began creating conventional infantry divisions complete with artillery support soon thereafter.

Strictly speaking, the French were not conducting COIN operations, but rather a reconquest of Vietnam. France had lost its authority over its colonies when the Japanese threw the French colonial administrators and their troops into jail. As noted above, Ho had declared the DRV in the interim and allowed the French to return only to replace the Nationalist Chinese overseeing surrender of Japanese forces north of the sixteenth parallel pursuant to the terms negotiated with the French in the March 6, 1946, Preliminary Convention. Under this agreement, France recognized the Republic of Vietnam as a "free state" within the Indochina Federation and French Union. The Vietnamese agreed to twenty-five thousand French troops for five years to relieve the Nationalist Chinese.[72]

There was also a domestic political dimension to France's Indochina war. It was deeply unpopular at home.[73] One of the consequences of this deep antipathy was that the government, in an attempt to manage public opinion, would not deploy those metropolitan military units manned by conscripts. The French Expeditionary Corps deployed the French Foreign Legion and colonial troops from all of France's protectorates and colonies: from Senegal, Tunisia, Morocco, Cambodia, and Laos, and volunteers from Thailand. In a sense, then, the French Expeditionary Corps in Vietnam constituted economy of force operations given the constraints on its access to French forces. As the historian Fredrik Logevall observed, "A growing number of officers thought there were simply not enough boots on the ground to stake everything on a military solution and little prospect that more could be found."[74]

Nevertheless, French military efforts focused initially on destroying the Viet Minh forces. In hindsight, this approach seems appropriate because ultimately what ordinary Vietnamese wanted did not matter. Although a majority of Vietnamese favored independence and a large sub-set of them embraced Ho Chi Minh and the Viet Minh, they were not the primary source of Ho's war-making support; Communist China and the Soviet Union were. That said, Ho "had captured the leadership of the Vietnamese nationalist movement and ha[d] a seemingly inexhaustible supply of recruits for his army."[75]

As the war unfolded, the French suffered a number of reversals. Emblematic of these was the case of Nghia La, attacked on October 17, 1952. The French lost seven hundred troops in defense of this outpost and, when it collapsed, cost France the entire line of outposts for which it had been the anchor.[76] Later in the autumn, Giap turned his attention to Black River French outposts, including Na San. Na San was unusual; it had not been one of the original outposts but had been built quickly about 117 miles from Hanoi in rugged terrain. Paratroops had seized the site, which was subsequently developed, fortified, and resupplied completely by air. The defensive complex featured a number of outposts within supporting range of each other, and an airstrip. General Giap's forces attacked Na San on November 23 and were badly mauled by the French defenders.[77] General Salan, the French commander, is said to have seen the potential of such bases to lure the enemy into combat, and then to inflict serious losses on them: "[after Na San] . . . in Salan's eyes, the theory of the air-ground base appeared to have been vindicated" and ultimately created the pattern in French operations that led to Dien Bien Phu.[78]

In trying to exploit the lessons they believed they learned at Na San, the French built a larger-scale but similar complex at Dien Bien Phu in the hopes of luring the Viet Minh to another mauling like the one meted out at Na San. If the French were able to destroy a sufficient number of Viet Minh forces, they hoped, the Vietnamese would have to negotiate a settlement. The location for the French air-ground base, however, was remote, out by the Laotian frontier; it was at the outer edge of the range that French air forces could sustain and support. Moreover, the French did not occupy the high ground around the base, which allowed the Viet Minh to emplace artillery there. Limited target location methods and dense vegetation on the heights prevented French artillery and air strikes from silencing the Vietnamese guns. The French garrison was subjected to steady bombardment while their supplies dwindled, their casualties grew, and Viet Minh infantry and engineers dug trenches to bring their assault forces ever closer to Dien Bien Phu's strong points, which collapsed one by one.

The French surrender at Dien Bien Phu was a major blow. As many authors have noted, it was the first time a European army had been defeated by its colonial subjects since the American Revolution. Logevall notes that it might not have been such a decisive event under different circumstances; France might have regrouped and soldiered on. Such an outcome was unlikely given the recent trauma of the Second World War and the political climate at home. Against this backdrop, the defeat took on additional significance. The government resigned and a new government led by Pierre Mendes-France supported French withdrawal from Indochina.

In contrast to Vietnam, the military problem confronting the French in Algeria was an insurgency. Moreover, Algeria was no colony, but rather a department (similar to a state) within the French Republic. Despite the fact that Algeria had become an integral part of France, the National Liberation Front (or FLN in its French acronym) enjoyed wide support for independence and received material support from Tunisia, Morocco, Egypt, the USSR, and Yugoslavia.[79]

According to historian Alistair Horne, the FLN did not pursue a strategy completely consistent with the revolutionary war theories of Mao and Truong Chin.[80] Instead, the FLN moved its leadership to sanctuaries in neighboring states, where it mistakenly thought they would be safe. Next, realizing that it was too weak to confront the French armed forces, it focused action on the native police. This concentration of effort forced the army to protect the police instead of pursuing search and destroy missions against FLN forces.[81] In addition to attacking the police, the FLN targeted Muslim moderates for death including soldiers and their families. Through these attacks, the FLN sought to remove "bridges of conciliation" that otherwise might have produced a settlement.[82] They also terrorized the rural population, whom the French had disarmed, which rendered this population mute as a source of information for French forces.[83] The FLN also undertook selected killings of Europeans, which had the effect of making all Muslims look dangerous and prompted some Europeans to move to town in search of greater security.[84]

By 1957, the FLN escalated its attacks to bombings. These bombs, often homemade, slightly larger than hand grenades, and equipped with a timer-detonator, were designed to be carried in beach bags and purses. The FLN recruited young Muslim women who wore European dresses and makeup and could often pass as European to penetrate French security and place bombs in cafes and other locations frequented by the French and the *pied noirs*, those born of European stock in Algeria.[85] These bombings undermined the general sense of security, especially in the larger cities. They also prompted the battle of Algiers during January through March 1957, in which the French routed the bomb-makers and FLN gunmen.

The French pursued a COIN campaign that can be summarized in the five-step efforts this chapter used earlier to capture the essence of British efforts in Malaya and Kenya.

Separate the Guerrillas from Their Support Bases

According to A. A. Cohen, biographer of David Galula, the French COIN expert, the first step was to isolate Algeria from outside interference. To this end, France deployed naval blockades to interdict arms and supply shipments inbound from Yugoslavia and elsewhere. The French also built barriers along

Algeria's frontiers with Morocco and Tunisia and deployed surveillance including "human and technical means" to seal off the southern deserts.[86] Within the populated areas of Algeria, the French instituted *quadrillage*, using checkpoints within cities to isolate neighborhoods, limit freedom of movement, and impose social controls on the inhabitants. Populated locales were assigned static forces to ensure security and protection. Mobile units, not permanently associated with any specific location, provided an offensive and reaction force capability. Collectively, the security architecture resulting from these several initiatives was intended to make movement to and from their support bases difficult for the guerrillas and to limit their freedom of action to carry out attacks.[87] Eventually the FLN fighters were either driven into the Atlas Mountains, where they were hunted by the mobile forces, or restricted to urban enclaves like the Kasbah in Algiers, where they were routed out by the 10th Parachute Division and its informants.

Impose Controls to Limit the Population's Ability to Support

The most important step in this regard was the increase to five hundred thousand troops following Prime Minister Guy Mollet's visit in February 1956.[88] The manpower increase meant France had twenty-three security forces for each one thousand of the Algerian population: a force-to-population ratio robust enough to exert positive control on civilian activities.[89] In January 1957 at the outset of the battle of Algiers, the French also imposed *quadrillage* on the city and its inhabitants, enhancing French forces' intelligence on FLN fighters in the Kasbah and elsewhere and creating a systematic dragnet to identify, locate, and capture the bomb-makers and gunmen. In addition, the French created local block wardens (*Dispositif de Protection Urbaine*) whose job it was to keep track of persons of interest in their area and report suspicious activities.[90] Though the wardens were understood by the FLN for the threat they posed and thus targeted for killing, they nevertheless provided valuable information to government forces and made it more difficult for FLN operators to hide themselves within the populace.

Offer People Something More Attractive Than the Guerrillas Are Offering

A. A. Cohen, citing French indigenous affairs officers, concluded that poor colonial administration was a major source of discontent driving the insurgency.[91] Horne notes that the French increased the Algerian department budget by 40 percent over the previous year to improve living standards there. In addition, the government decreed that 50 percent of all vacancies in public service would be reserved for Muslims.[92] These inducements could not compete with the FLN promise of independence, however. When President de Gaulle conducted a referendum in May 1961 on the question of independence from France, the response was an overwhelming "yes" for independence.[93]

Create Rewards and Incentives to Support the Government and Quit the Guerrillas

"The vile hand of torture"[94] served as an incentive to avoid contact with the FLN; those who were implicated often suffered brutal interrogations. It also worked in the opposite direction; perceived cooperation with the French authorities could lead to equally brutal FLN reprisals. Positive incentives to support the government were few, but the Constantine Plan provides one example. It created one thousand villages and a vigorous agricultural support program that increased agricultural production from 270,000 tons to 390,000 tons and made other improvements in Muslim lives.[95] The French incentives and hearts and minds programs appear to have enjoyed the greatest support among the indigenous affairs personnel and COIN theorists such as Galula. The centerpiece of the French campaign plan, however, remained the destruction of enemy forces.

Destroy the Enemy

After the battle of Algiers, the breaking of the FLN-directed general strike, and successful actions to decapitate the FLN's leadership, many senior French officers believed they had beaten the insurgents. Indeed, French paratroopers had found the bomb-makers and their laboratories, and removed them as the FLN's principal weapon. The French forces had also broken the general strike by tearing the shutters off shop fronts and exposing them to looting, quickly compelling the owners to return to protect their investments. The troopers also rounded up school children and drove them to school. Perhaps most impressively, the French had captured Ahmed Ben Bella, one of nine members of the Revolutionary Committee for Unity and Action, on a flight out of Morocco in 1956.[96]

Yet it was the inability to run the FLN to ground in Tunisia and Morocco that, according to Alistair Horne, led to the collapse of the Fourth Republic and the advent of de Gaulle in 1958.[97] With de Gaulle's return to power in the Fifth Republic, the consensus on Algeria began to fray. De Gaulle changed positions on Algeria between his 1958 commitment to a French Algeria and support for an Algerian Algeria in 1960. The general's shift in policy position outraged some of the senior leadership in the army, who, feeling betrayed, spawned the secret army organization OAS, which took it upon itself to maintain French control of Algeria. The generals led a putsch meant to intimidate Paris into a change in policy. As a part of this effort, a regiment of paratroopers conducted a parachute assault on Corsica and threatened a similar assault on Paris. De Gaulle's support for Algerian independence put the *pied noir* community in a dangerous position where they believed they must flee or face retribution from the Muslim majority: "the suitcase or the coffin," as the saying went.

It may have been that the French military forces had succeeded in defeating the FLN in the field, but had taken too long to accomplish it, and left the FLN political agenda intact, especially the demand for independence. Moreover, seven years of war created an exhausted population, both in Algeria and in metropolitan France. "As much as any other factors, it was this exhaustion that helped bring the civil war to an end."[98]

Applicability of Colonial Lessons to U.S. Circumstances

In the context of the Cold War, the canonical British and French examples of COIN may have been apt; they certainly provided rich catalogs of tactics, techniques, and procedures. General Salan argued during a visit to Washington, DC, that the Indochina war he was waging and the ongoing Korean war were equivalents in the battle to contain Communism. The United States appears to have agreed with this characterization—up to a point—increasing its military assistance to France following the general's visit. The United States, however, never adopted in full the practices of either the British or the French. In Vietnam, Iraq, and Afghanistan, U.S. forces employed some of the simpler tactics, techniques, and procedures from the French and British playbooks but never attempted the more difficult—and arguably essential—tasks such as population movement controls, food supply controls, and large-scale population relocation efforts.

As a result, however salient these British and French cases once were, they are less so now. There are several reasons.

The United States Is Not an Imperial Power

The United States is not an imperial power and therefore has limited authorities and limited capabilities for COIN as practiced by the British and French. The United States has typically sought to employ its COIN capabilities to assist a friendly government confronting an armed opposition. In this role, U.S. forces must work with the local authorities to formulate and implement COIN policies and actions. The United States could not, on its own authority, relocate large segments of the host nation's population as the British did in Malaya and Kenya. Nor could the United States insist on a policy of *quadrillage*, as the French did in Algeria. Even though it has often been able to coerce the partner government into preferred courses of action, Washington has ultimately needed the partner's consent, however begrudgingly it was given.

Because the United States is not an imperial power, it has not built the capabilities necessary for COIN in a colony. The U.S. government has no capability for foreign, civil administration (as both retired Lt. General Jay Garner and his successor, L. Paul Bremer, demonstrated in Iraq when they tried to oversee the ad hoc Coalition Provisional Authority) and no police force or Special Branch–like capabilities (military police fulfill a narrower role). The absence of these capabilities contributes to the United States' difficulties in Phase IV operations, where stability and COIN actions typically take place.

Because the United States works "by, with and through"[99] local authorities, it must have a local government with which to partner. Often, the choices are few: the Diem brothers in South Vietnam, who were remote from the people and consumed by conspiracy theories; the Karzai government in Afghanistan, wallowing in corruption and resisting the United States at every turn; and the Maliki regime in Iraq, promoting factionalism and attempting to marginalize the Sunni population whenever possible. This imperative for a local partner has in recent years shackled U.S. policy to a series of autocrats interested in ruling more for the material benefits of the job than out of a sense of duty to improve the lives of their people.

British and French Approaches Worked Differently Than They Claimed

Eric Stahl, then an undergraduate at Tufts University, published an honors thesis making the case that the deck was stacked against the Communists from the beginning.[100] The Communist Party of Malaya (CPM) made decisions not to pursue alliances and cooperation with other independence-minded groups during World War II. They also passed up opportunities to collaborate with the Japanese. As a result, the party commanded a relatively small band of CTs and supporters. The forces the CPM did command were woefully underequipped with virtually every staple of armed resistance: arms, ammunition, communications equipment, transportation, and food. The insurgent force was so small and dispersed it could not take advantage of British incompetence early in the war.[101] Ultimately, according to Stahl, the emergency was lost by the CPM more than won by the British.

Similar circumstances doomed the Mau Mau rebellion. The British colonial authority in Kenya was brutal and racist, and it took sweeping action against the Mau Mau, who were pursuing a peasants' revolt.[102] As this chapter described earlier, the revolt was deprived of its leadership from the beginning by ruthless British action. This left the peasants, the ordinary Kikuyu, with aspirations for independence but no plan or resources. Their fate was only a matter of time.

The French military claim, that it had inadequate forces on the ground in Vietnam, was certainly true, but other factors contributed to the Expeditionary Corps' defeat. Dien Bien Phu was not simply a case of military planning gone wrong; it was a reflection of the Viet Minh's growing military capabilities. Indeed, after the surrender at Dien Bien Phu, the Viet Minh visited another military disaster—albeit on a somewhat smaller scale—on the French with the ambush and destruction of *Groupement Mobile 100* in the Mang Yang Pass down in the Central Highlands of the country.

Moreover, French military manpower was so limited that the French command basically forfeited control of the southern part of the country to the Viet Minh after the initial pacification effort. As one of the Legionnaires remarked, French control in the south extended only seventy-five meters on either side of the highways.

In the case of Algeria, President de Gaulle gradually came to understand the futility of the military effort. Algeria enjoyed strong third-party support, not only from its neighbors, Tunisia and Morocco, but also from Yugoslavia and the Soviet Union. French efforts to isolate Algeria from its outside support had been ineffective. The victory in the battle of Algiers was largely tactical. If de Gaulle had not ended the war with a grant of independence, the FLN could have fought on indefinitely—as de Gaulle realized—with potentially grave consequences for the fragile Fifth Republic.

The Societies and Cultures Subject to Insurgency Are Different Today

The insurgencies the United States confronts in the current era are distinct from the wars of national liberation with Marxist-Leninist overtones of the Cold War. For one thing, today's partners and their adversaries are not simply the poor, agrarian societies plagued by insurgency as in earlier times, but more complex entities. They are not striving for independence from the colonial Metropole and are rarely motivated by notions from Marxism-Leninism. Today, U.S. forces must operate in cultures where pursuit of Koranic righteousness often trumps the appeal of secular democracy and material well-being. Digging wells, painting schools, and similar developmental projects—the repertoire of civic action teams in earlier insurgencies—are less efficacious in these populations. There may sometimes be instances, like the al Anbar province awakening, where the logic of "the enemy of my enemy is my friend" combined with improved security provided by U.S. Marines may have given U.S. forces temporary traction,[103] but such cases are rare. These societies are complex, embodying multiple tiers of tribal, sectarian, technical, and governmental elite expectations that the U.S. policy community almost never understands and therefore rarely addresses.[104] States like Pakistan are simultaneously poor, wild, and clannish, armed with nuclear weapons and capable of stymying many U.S. policy demands. Iraq is simultaneously an ancient entity in the heart of the old Caliphate and a modern, failed secular state largely incapable of defending its mixed polity against ISIS or exerting its own authority throughout its territory—despite the massive investments Washington has made in support of these objectives. Despite its poverty, Afghanistan has successfully resisted Alexander the Great, the British, the Soviets, and U.S.-led coalition efforts to recast it, respectively, as a Hellenistic state, colony, Marxist-Leninist puppet state, and modern democratic state.

As the United States observes without fully understanding unfolding events, U.S. officials are surprised when seemingly sound allies like Yemen suddenly succumb to Houthi violence even as U.S. special operations forces partner with Yemeni forces for training and military cooperation. The advent of ISIS in Iraq, the negative consequences of the Arab Spring, the rampant propagation of militias, and growing chaos across the CENTCOM area of responsibility all seem to outpace the United States' ability to comprehend and respond in ways consistent with our national interests. These developments are not treatable with the COIN tactical playbook of train, advise, and equip;[105] success is less about providing good government and delivery of social services and more about satisfying profound religious and cultural urges—or constraining them.

Today's Insurgents Reflect Two Main Types

Of course, some regions continue to produce old-fashioned insurgencies where the enemy combatants are indigenous fighters seeking to oust the government. The insurgents the United States is most likely to encounter are different. Today, the modern insurgents are increasingly part of a state's military instrument. General Giap's army in Vietnam gave us an early glimpse of such a force. His army by 1954 had six regular divisions and hundreds of regional and local forces. All of these forces were part of a single military instrument, executing a single strategic plan, although each of these groups of forces had unique roles to play in executing the strategy.

The modern manifestation of this approach is Iran. Today, Iran has a traditional, regular set of armed forces, but they are complemented by a series of paramilitary and special operations forces (e.g., the Quds Force) and regionally grown militias. Tehran supports the Houthis in Yemen, whose militias respond to Iranian instructions. Hezbollah in a similar way represents the tip of Tehran's spear in Lebanon. In Syria, the shabiha militias operate as an extension of the Assad regime's regular armed forces, serving as augmentation to government forces efforts in the ongoing Syrian insurrection.

The second type of insurgent force is the tribal or sectarian militia. In Iraq, the Shia community has spawned multiple militias as the Iraqi army and police formations have shown themselves incapable of defending against ISIS. Militias as insurgent forces are historically unusual in the sense that insurgencies usually begin as actions by the few: a revolutionary-minded figure (e.g., V. I. Lenin) or a small, dispossessed group (e.g., the Malayan Chinese minority). The tribal and sectarian militias appearing today are not minority affairs; they represent and seek to protect an entire constituency such as Iraqi Shia or Alawite Syrians. They exist because the official regime has failed them.

Suitability of the Doctrinal Solutions Prescribed in FM 3-24

If, for the reasons just presented, the canonical cases do not present sound templates for U.S. COIN operations, what about current U.S. Army and Marine Corps doctrine? Field Manual 3-24 presents two main approaches: an indirect approach and a direct approach.[106] The indirect approach emphasizes national assistance, security cooperation, generational engagement, negotiation diplomacy, and a process to identify, separate, isolate, and influence the insurgents, and eventually, to reintegrate them into society. The direct approach concentrates on a process to shape the operating environment, clear areas of insurgent activity, hold those areas while building up their infrastructure and social services, and ultimately transitioning responsibility back into the hands of local authorities.[107]

James F. Jeffrey, U.S. ambassador to Iraq from 2010 to 2012, offered a critique of U.S. strategy noting, "the blame [for U.S. counter-insurgency] lies not with poor implementation but with the strategy itself."[108] In particular, Ambassador Jeffrey identified the trouble associated with "build"-related activities, stating, "the deck is stacked against success." He went on to note other weaknesses of the U.S. approach:

> U.S. ground troops as opposed to military advisors or air forces, almost always generate negative reactions among developing world populations. [In addition,]

neighboring states, alarmed by the prospect of U.S. forces on their borders, often have an interest in supporting the insurgents. North Vietnam and China did so in South Vietnam, Pakistan did so in Afghanistan and Iran and Syria did so in Iraq.[109]

RAND senior analyst James T. Quinlivan published an analysis of force requirements for stability operations that examined the force-to-population ratios associated with various stability/COIN operations.[110] In general terms, the higher the force-to-population ratios, the more likely they are to produce positive outcomes. For example, Quinlivan noted that ratios of approximately twenty or more security forces per thousand of population were required for environments like that found in Northern Ireland during the Troubles. In contrast, the United States was able to deploy approximately 0.2 security forces per thousand in Iraq in 2004. In Afghanistan, if the Northern Alliance troop strength is included in the computation, the United States managed 0.7 forces per thousand of population. These meager ratios suggest one of many reasons why the United States failed in both cases. Its armed forces were too small to generate the required force-to-population ratios that are prerequisites for success, and Washington was unwilling or unable to raise indigenous forces to improve the ratios. Thin force-to-population ratios mean that the COIN force lacks the manpower to implement the direct approach doctrine (e.g., shape, clear, hold, build, transfer) found in FM 3-24.

The indirect approach to COIN presented in the Field Manual is only practicable when the beleaguered state still has armed forces that can benefit from security cooperation and similar programs to counter the guerrilla forces. The United States' experience in South Vietnam, Iraq, and elsewhere, in which the U.S. invested a fortune in blood and treasure and over a decade without producing local armies capable of defeating the adversary (or developing a local economy able to sustain such an army), suggests that the indirect approach is not a viable policy option for confronting future insurgencies.

RAND political scientist Christopher Paul and his colleagues have conducted a large body of research that bears on the prospects for success using current U.S. COIN doctrine. First, Paul notes that the odds are long for the COIN force. In a study of thirty insurgencies between 1978 and 2008, COIN forces prevailed in eight of the thirty examples: approximately 27 percent.[111] In a later study, Paul and his team found "three COIN concepts present in each and every COIN win and (not implemented in their entirety by any losing) COIN force: tangible support reduction (e.g., border controls), commitment and motivation (government and COIN force), and flexibility and adaptability (COIN force)."[112] The later study also noted that "no external COIN force or externally supported COIN force was able to prevail if the host nation government was insufficiently committed."[113] Finally, Paul noted that the median length of time for an insurgency in seventy-one cases was 118 months: a little less than ten years. Insurgencies in which COIN forces prevailed took longer, with a median length of 132 months.[114] Long-duration COIN fights are not good for all-volunteer armies, as the years in Afghanistan and Iraq—and the resulting post-traumatic stress disorder and suicide casualties—have made clear.

Paul's research is not good news for the United States. The host nation governments that Washington supports rarely appear capable and committed to the fight. The borders, whether in Vietnam, Afghanistan, or Iraq, are vast and porous, making it nearly impossible to prevent tangible outside support. And U.S. forces are not very flexible. They often make minor adjustments to their tactics, techniques, and procedures, but above the tactical level, little changes. Decisions focus on troop strength; hence, the "surge" in Iraq stood out as remarkable. The sort of flexibility that would exploit success against the adversary or capitalize on the enemy's demonstrated weaknesses is in short supply.

Conclusions and Observations

In the near-term, the insurgencies likely to prompt U.S. military intervention will most probably occur in the CENTCOM area of responsibility where access to oil, freedom of navigation, and obligations to allies (e.g., Turkey, Saudi Arabia, Bahrain) collide with radical Islam and the collective aspirations that spawned the Arab Spring and that seek to escape the constraints imposed by long-standing autocratic regimes. Some insurgencies may be outgrowths of today's wars in Syria and Iraq and elsewhere, but fate always leaves room for new, unanticipated conflicts to surprise us. If events reveal new political entities emerging to exert authority over the remnants of Iraq and Syria, perhaps these fledgling governments could become the committed, capable partners that Christopher Paul's research found to be central to COIN success.

If the United States expects to be successful in future COINs, it must be able to deploy a force structure suitable for all of the tasks across all phases of operations, from combat through stability and hand-off to local authorities. In particular, it must have military governance units that can provide order and security and also deliver the full range of civic services including courts, police, public utilities, and emergency services. And these units must be abundant enough within the force structure to sustain operations until responsibility can be transferred to local authorities, the United Nations, or some other international organization (e.g., Organization of African Unity, Arab League). These units must also be robust enough so that in worst-case circumstances, they can provide the service or function in question directly (e.g., operate public utilities). In many instances, they may only serve as advisors and trainers to improve the performance of local officials.

If the United States embarks on future COIN operations, it must be clear about the defeat mechanisms it will pursue. These are, first and foremost, the destruction of the insurgency's forces, and second, elimination of its outside support. Given the limitations on the size of forces the United States can deploy on a sustained basis, the modest size of forces contributed by allies and partners, and the limitations, in both size and quality of indigenous forces, the United States must be able to substitute technology for manpower to achieve controls over population and territory equivalent to robust force-to-population and force-to-space ratios. Moreover, as Christopher Paul's research shows, winning COINs typically takes years—years of engagement that are nearly impossible to sustain with a small, all-volunteer ground force. The system of instrumenting the population and its environs and imposition of social controls described here offers the United States an alternative approach that may let its forces escape the fate of initial success followed by a grinding war of attrition by accelerating the rate at which insurgents are eliminated and compelling states to cease their support for the insurgency.

Destroying the Enemy's Forces

The first step toward destruction of the insurgent forces is the ability to identify and locate them in a foreign society whose many dimensions are opaque to U.S. observers. To address this limitation, the United States should instrument the major towns, cities, and lines of communications of the country and its population with surveillance technology. Priority should go to those regions experiencing the greatest violence. Industrial supply chains and their contents have been instrumented in this way, giving reason to believe that a society could be as well, albeit with more limited efficacy. In addition to deploying the usual surveillance and reconnaissance systems, instrumentation involves issuing everyone identity cards that contain a radio frequency identification (RFID) chip and covering the landscape—urban and rural—with active interrogators (i.e., chip readers) that will allow authorities to track the movements of everyone in the population. Public buildings, road intersections, bridges, market places, and even private homes could be instrumented. Anyone found without his or her identity card would be detained and questioned. Cars, trucks, boats, and other forms of conveyance along with livestock would all be chipped and tracked. Vehicles entering the country would be chipped as well. RFID readers would be supplemented with camera surveillance.

Next, a program of social controls would be implemented. It would include a curfew and strict controls on access to food, fuels, fertilizer, explosives, weapons, and computers and telecommunications equipment. Freedom of movement would be limited, allowing people movement only between their homes, workplaces, schools, and markets, where many of the controlled goods would be available in rationed amounts. Commercial traffic would require permits and would have to use assigned routes. Commercial telecommunications would be shut down or jammed. Cell phones would be illegal and confiscated.

The instrumentation of society would provide the authorities with information that could be turned into a template for normal, everyday life (e.g., the authorized movements from home to work or school) that would make identification of deviations from the normal routine possible. When attacks such as car bombs take place, the authorities would be able to identify those present at the location immediately prior to the attack and bring them in for questioning. The combination of cameras and RFID readers would also make it possible to identify someone who is seen to be present but provides no response to the nearest RFID reader, and who can then be tracked on video so the police can make an arrest.

This system would be imperfect, and the enemy would find some methods for subverting it. Nevertheless, it would be a major improvement for U.S. surveillance and intelligence capabilities and would simplify what security forces need to understand. Rather than finding themselves confronted with a blizzard of factoids about tribal relations, social hierarchy, and the like, all the authorities need to know is whose presence consistently correlates with acts of violence. Thus identified, suspects can be visited at their homes, arrested, and questioned without violating the privacy of entire neighborhoods. Drones equipped with RFID interrogators and personnel sensors could patrol remote areas. When specialized sensors clearly indicate human presence but the RFID reader is silent or receives relatively few responses given the number of people indicated by the other sensors, the site can be identified as an enemy camp or safe house. Appropriate long-range strikes could be authorized, or, alternatively, other intelligence collection means (e.g., reconnaissance patrols, listening posts) could be brought in range to monitor the site and collect potentially useful information from it.

Elimination of Outside Support

Outside support to insurgents from neighbor states is difficult to stop, especially when the shared borders are long and the terrain is difficult and therefore affords good opportunities for camouflage, cover, and concealment. It is not impossible, however, and the United States has options. In some instances, a stern note of protest may be enough. News reports indicate that Iran was convinced to stop supplying explosively formed penetrators (EFPs—simple devices that amplify and focus the effects of improvised explosive devices

[IEDs]) to Iraqi combatants by threats from Washington.[115] Punitive strikes are another option, in which the United States would destroy critical infrastructure or similar high-value assets to coerce a supportive state or similar entity into curtailing its assistance to the insurgents. The premise is that if the costs of the damage exceed the benefits expected to accrue from support to the insurgency, the support will diminish or stop. Imagine cruise missiles raining down on facilities crucial to the Iranian nuclear program or its energy industry. Lesser punitive operations might mine ports or disrupt national telecommunications. A defensive counter-air campaign could kill the Iranian air force on the ground and leave Iran vulnerable to subsequent attacks. The Quds Force itself, which often provides the military advice and assistance to client insurgencies, could be targeted. The nation's money and banking systems might be disrupted through cyber-attacks. Seventy-seven million Iranians unable to cash a check as a result might give the leadership in Tehran pause.

There are also border controls that could be implemented. Land border crossings could be fortified and closed to traffic, stopping trade. The border region could be depopulated and turned into a free fire zone, sewn with mines and patrolled by drones and manned aircraft. Air traffic could be restricted to specific corridors in a limited number of airports. In such an environment, cargo aircraft bound for insurgent camps would be more easily detected and destroyed. Maritime traffic could be patrolled and boarded to establish the nature of the cargo and its destination.

Of course the barricade will leak and some support will find its way to insurgent forces. Of course the supporting entity will claim innocence and seek covert means through which it can continue its support while plotting retaliatory actions. But if the insurgents can be deprived of their ability to move, communicate, and plan operations by the social controls described above, and if these same social controls deprive the insurgents of food, shelter, and material support from the local economy, their operational capabilities will be substantially undermined and remaining forces will fall prey to U.S. operations able to identify and locate them.

Likewise, if the costs associated with providing outside support are large enough and impose penalties the supporting entity does not want to endure, then support to the insurgency will likely decline. Moreover, as U.S. and partner forces succeed in killing off the insurgents (or they begin surrendering), at some point they will lose critical mass and no longer be capable of conducting operations. Somewhere along this trajectory, the supporting entity is likely to conclude that there is nothing left of the insurgency worth supporting.

Among the virtues of the proposed approach is the fact that it makes no pretense of understanding root causes of the insurgency. It simply suggests a method for focusing the application of military force based on correlations between individuals' presence with acts of violence. It focuses on things security forces can do well, such as finding and destroying the enemy, without encumbering the U.S. and partner forces with tasks they do less well, such as nation building and winning hearts and minds.

The prescription for treating states and non-state actors that support insurgencies may seem risky, but this approach has worked with varying degrees of success in the past for similar purposes. As president, Thomas Jefferson used punitive naval actions against the Barbary pirates in 1804 to dissuade them from plundering U.S.-flagged ships and enslaving U.S. sailors. The United States occupied the Mexican city of Veracruz in 1914 to force the Mexican government to abandon policies of which the United States did not approve, particularly the habit of harassing and jailing U.S. sailors on shore leave in the city, and to prevent a shipment of guns from sailing for Germany. The United States carried out punitive strikes against North Vietnam to punish it and attempt to dissuade it from continuing its aggression against the Republic of South Vietnam. The Clinton administration carried out punitive strikes in response to the first attack on the World Trade Center and for the attacks on U.S. embassies in Tanzania and Kenya.

Punitive actions are likely to remain an important option in the future. Where the state to be punished is nuclear-armed, there are certainly reasons to proceed cautiously, but caution should not lead automatically to paralysis. Armed confrontations have occurred between nuclear-capable states in the past without escalating to nuclear strikes. The Chinese-Soviet confrontation along the Ussuri River in 1969 de-escalated when Chinese emissaries flew to Moscow. Confrontations between India and Pakistan over Pakistani-based terrorism from 1990 to 2008 also concluded short of nuclear war. Turkey's shoot-down of a Russian Sukhoi SU-24 attack aircraft in November 2015 produced outrage from Moscow but no direct, military response. These episodes suggest punitive strikes remain a viable option for dealing with states that support insurgencies (or terrorism), although the specifics of the actions must be carefully analyzed.

Just as the United States dominated the countries it coerced with punitive actions in the past, it can dominate the likely suspects in CENTCOM's area of responsibility in the near-term future. The strongest of these, Iran, has a crippled economy from years of Western sanctions (although those sanctions are vanishing with the nuclear treaty that entered into effect in January 2016), an obsolescing conventional military establishment, and a collection of irregular or paramilitary formations of uncertain capabilities. Iran also possesses a stockpile of short- and intermediate-range missiles, but, insofar as anyone in the West knows, they are not yet nuclear-capable. The United

States has the means to dominate Iran at every step on the escalation ladder on the use of military force. Washington should enjoy significant freedom of action in punishing Iran or any state in the region that supports insurgency operations against U.S. interests.

This chapter has described some draconian steps for establishing security and safety in stability operations and COIN actions. They are bound to raise questions about whether they only create more terrorists. None of these actions should be adopted automatically, and outcomes may vary depending upon specific circumstances. In any case, the president and his senior defense advisors must weigh the consequences of losing an otherwise successful campaign in the end game of Phase IV operations and what those consequences mean to America's security and other vital interests. Ideally, these should be matters considered carefully in the earliest deliberations by the president and his advisors about how the United States should deal with an unfolding crisis—and well before forces deploy.

Notes

1. General Sir Rupert Smith, *Utility of Force: The Art of War in the Modern World* (London: Penguin, 2005).
2. "The guerrilla must move amongst the people as a fish swims in the sea." Cited at http://www.brainyquote.com/quotes/quotes/m/maozedong138236.html#qoUkszj83VGb2mLh.99.
3. Alistair Horne, *A Savage War of Peace: Algeria 1954–1962* (New York: New York Review Books, 1977), and A. A. Cohen, *Galula: The Life and Writings of the French Officer Who Defined the Art of Counterinsurgency* (New York: Praeger, 2012).
4. Max Boot, *Invisible Armies: An Epic History of Guerrilla Warfare from Ancient Times to the Present* (New York: W. W. Norton, 2013).
5. See for a concise description, Phillip Deery, "Malaya, 1948: Britain's Asian Cold War?" *Journal of Cold War Studies* 9, no. 1 (2007): 29–54.
6. Noel Barber, *The War of the Running Dogs* (London: Cassell, 2004), 247.
7. Ibid., 38.
8. Ibid.
9. Wunyabari O. Maloba describes the rebellion as a classic peasant uprising. See Maloba, *Mau Mau and Kenya: An Analysis of a Peasant Revolt* (Bloomington: Indiana University Press, 1998), 4.
10. Daniel Branch, *Defeating Mau Mau, Creating Kenya: Counterinsurgency, Civil War, and Decolonization* (Cambridge: Cambridge University Press, 2009), 6.
11. Ibid., 57.
12. Ibid., 58.
13. Barber, *The War of the Running Dogs*, 78.
14. Ibid., 98.
15. Ibid., 111.

16. Ibid., 112.
17. Ibid., 116, 212.
18. Branch, *Defeating Mau Mau*, 6, 7.
19. Some of these actions included torture, castration, and rape. In 2013, the British government paid £14 million to surviving Mau Mau members who had endured mistreatment at the hands of colonial officials. See Matt Chorley and Leon Watson, "Mau Mau Fighters Raped, Castrated and Beaten in Kenya's Uprising against Britain get £14 million Compensation But No Apology," *Daily Mail*, June 5, 2013, http://www.dailymail.co.uk/news/article-2336418/Mau-Mau-fighters-raped-castrated-beaten-Kenyas-uprising-Britain-14million-compensation-apology.html.
20. Branch, *Defeating Mau Mau*, 61.
21. Ibid., 89.
22. Ibid., 104, 105.
23. Barber, *The War of the Running Dogs*, 131.
24. Ibid., 83–85.
25. John Coates, *Suppressing Insurgency: An Analysis of the Malayan Emergency, 1948–1954* (Boulder, CO: Westview Press, 1992), 74.
26. Barber, *The War of the Running Dogs*, 131.
27. Ibid., 163, 321.
28. Ibid., 191.
29. Frank Furedi, *Colonial Wars and the Politics of Third World Nationalism* (London: I.B. Tauris, 1996), 180.
30. Branch, *Defeating Mau Mau*, 61–70.
31. Ibid., 72.
32. Ibid., 107, 111.
33. Ibid., 104.
34. Barber, *The War of the Running Dogs*, 80.
35. Kumar Ramakrishna, "'Telling the Simple People the Truth': The Role of Strategic Propaganda in the Malayan Emergency," *Journal of the Malaysian Branch of the Royal Asiatic Society* 75, no. 1 (2002): 49–68.
36. Barber, *The War of the Running Dogs*, 103.
37. Ibid., 117.
38. Ibid., 129.
39. Ibid., 182.
40. Ibid., 210.
41. Ibid., 229.
42. Ibid., 243, 244.
43. Ibid., 160.
44. Branch, *Defeating Mau Mau*, 121.
45. Ibid., 106.
46. Ibid., 121.
47. Ibid., 122.
48. Ibid., 130.
49. Ibid., 92.
50. Barber, *The War of the Running Dogs*, 69.
51. Ibid., 84.

52. Kumar Ramakrishna, "'Bribing the Reds to Give Up': Rewards Policy in the Malayan Emergency," *War in History* 9, no. 3 (2002): 332–53.
53. Barber, *The War of the Running Dogs*, 134.
54. Ibid., 144.
55. Ibid., 188.
56. Ibid., 234.
57. Ibid., 320.
58. Branch, *Defeating Mau Mau*, 86.
59. Ibid., 118–29.
60. Ibid., 148.
61. The Malaysian government continues to pursue reparations for the killings. See http://www.colonialismreparation.org/en/apologies-and-reparations-for-batang-kali.html.
62. Barber, *The War of the Running Dogs*, 66.
63. Ibid., 248.
64. Ibid., 320.
65. Ibid., 319.
66. Branch, *Defeating Mau Mau*, 67. Huw Bennett emphasizes the coercive side of the British approach. See Bennett, *Fighting the Mau Mau: The British Army and Counterinsurgency in the Kenya Emergency* (Cambridge: Cambridge Military Histories, 2012).
67. Branch, *Defeating Mau Mau*, 91.
68. Ibid., 5.
69. Ibid., 117.
70. Ibid., 5.
71. http://www.historyplace.com/unitedstates/vietnam/index-1945.html.
72. Fredrik Logevall, *Embers of War: The Fall of an Empire and the Making of America's Vietnam* (New York: Random House, 2012), 133.
73. Martin Windrow, *The Last Valley: Dien Bien Phu and the French Defeat in Vietnam* (Cambridge, MA: Da Capo Press, 2004), 48.
74. Logevall, *Embers of War*, 136.
75. Ibid., xii.
76. Ibid., 323.
77. Ibid., 328.
78. Ibid., 329.
79. Cohen, *Galula*, 131.
80. Truong Chin, *The Resistance Will Win* (pamphlet).
81. Horne, *A Savage War of Peace*, 16.
82. Ibid., 136.
83. Ibid., 134.
84. Ibid., 135.
85. Ibid., 184, 185.
86. Cohen, *Galula*, 136.
87. Ibid.
88. Horne, *A Savage War of Peace*, 150.

89. James T. Quinlivan, "Force Requirements for Stability Operations," originally published in *Parameters* (1996), http://www.rand.org/pubs/reprints/RP479.html.

90. Horne, *A Savage War of Peace*, 190, 198.

91. Cohen, *Galula*, 163.

92. Horne, *A Savage War of Peace*, 155.

93. Référendum sur l'autodétermination en Algérie, http://mjp.univ-perp.fr/france/ref1961.htm.

94. Horne, *A Savage War of Peace*, 18.

95. Ibid., 421.

96. Ibid., 184–91, 194.

97. Ibid., 18.

98. Ibid., 16.

99. This characterization of the U.S. approach comes from U.S. Army Field Manual 3-07.1, Security Force Assistance.

100. Eric Stahl, "Doomed from the Start: A New Perspective on the Malayan Emergency," 2009. Available from the Tufts University website as UA005.001.0064.000001. archival.

101. Chin Peng acknowledged these points in his book, *My Side of History* (Singapore: Media Masters, 2003).

102. Maloba, *Mau Mau and Kenya*.

103. Gabriel Ledeen, "Re-creating Anbar's Awakening," *Huffington Post*, April 27, 2009, http://www.huffingtonpost.com/gabriel-ledeen/re--creating-anbars-awake_b_179589.html.

104. Emma Sky, *The Unraveling: High Hopes and Missed Opportunities in Iraq* (New York: Public Affairs, 2015), chapters 19 and 20.

105. The train, advise, and equip efforts of the DoD seem problem-ridden. See the DoD Inspector General's report, DODIG 2015-093, Summary of Lessons Learned—DoD IG Assessment Oversight of "Train, Advise, Assist, and Equip" Operations by U.S. and Coalition Forces in Iraq and Afghanistan, http://www.dodig.mil/pubs/documents/DODIG-2015-093.pdf.

106. Headquarters, Department of the Army, Field Manual 3-24, Table 1-1.

107. FM 3-24, Table 1-1.

108. James F. Jeffrey, "Why Counterinsurgency Doesn't Work," *Foreign Affairs* (March/April 2015): 178–80.

109. Ibid.

110. Quinlivan, "Force Requirements for Stability Operations."

111. Christopher Paul et al., *Victory Has a Thousand Fathers* (Santa Monica, CA: RAND Corporation, 2010).

112. Christopher Paul et al., *Paths to Victory: Lessons from Modern Insurgencies* (Santa Monica, CA: RAND Corporation, 2013), xxiii.

113. Ibid., xxix.

114. Ibid., xxxi.

115. See Andrew deGrandpre and Andrew Tilghman, "Iran Linked to Deaths of 500 U.S. Troops in Iraq, Afghanistan," *Military Times*, July 15, 2015, http://www.militarytimes.com/story/military/capitol-hill/2015/07/14/iran-linked-to-deaths-of-500-us-troops-in-iraq-afghanistan/30131097/.

5

Insights on the Use of Military Force

THE UNITED STATES HAS AMASSED A POOR RECORD in military operations since the end of the Second World War: a record that deteriorated further after the end of the Cold War. In this latter era, it has become popular in military circles to claim that the armed forces did their part, but other agencies fumbled the ball, snatching defeat from the jaws of victory. Surely other agencies, especially those attempting to run the post-cease-fire interim governments in places like Iraq, have demonstrated their shortcomings. But I believe a future U.S. military, appropriately trained and led, responding to the president's guidance, should be able to deliver outcomes closer to what the president asked for than we have seen in the recent past.

This chapter presents lessons and insights from U.S. military experience in support of this claim.

Lessons from the Second World War

An important lesson from the Second World War is to realize how much destruction was required to kill the Nazi regime and to bring the Imperial Japanese government to its knees. In Europe, the Soviet Red Army destroyed more than three Wehrmacht army groups and seized Berlin. The Red Army also proceeded systematically to brutalize the German population. In Berlin alone, between 10 and 20 percent of the women were raped.[1] On the western front, American and allied forces killed their way through France, the BENELUX countries, and into Germany as well. Moreover, before the Western allies ever landed on the Normandy beaches, the allied air forces had bombed German industrial cities to rubble. It was Eisenhower's threat, made to German General Alfred Jodl in the last days of the war, to resume bombing of Germany's cities that finally produced complete capitulation.[2] A total of 4.2 million Germans, military and civilians, died during the course of the war.[3]

In Japan, well before the atom bombs detonated above Hiroshima and Nagasaki, allied air forces had firebombed Japan's most populous cities. Much of the Imperial Army had been destroyed on islands across the Pacific or found its units stranded in garrisons far from home. A total of 3.25 million Japanese died during the war.[4]

In both instances, the German and Japanese people had no illusions about their circumstances.[5] They ultimately understood that they had been defeated. Under occupation by allied forces, the victors killed another 2.5 million Germans in the war's aftermath.[6] Perhaps especially in the Red Army zone of occupation, the population was prepared to endure nearly any deprivation rather than have the killing and destruction of the deadliest days of the war grind on. And this despite the fact that Admiral Karl Doenitz, the heir to Hitler's regime, still presided over a formidable force in Norway, Denmark, and northern Germany of some four hundred thousand troops.[7] The scale of the destruction and death, combined by the allied powers' large forces able to swiftly occupy enemy territory and impose their authority on it, prevented any organized resistance from challenging them. As Doenitz explained in his memoir,

> The picture of the military situation as a whole showed clearly that the war was lost. As there was also no possibility of effecting any improvement in Germany's overall position by political means, the only conclusion to which I, as head of state, could come was that the war must be brought to an end as quickly as possible in order to prevent further bloodshed.[8]

In the cases of Afghanistan and Iraq, the United States and its partners never achieved similar conditions. The scale of death and destruction was relatively modest. Iraqi soldiers simply abandoned their uniforms and went home. Once the initial "shock and awe" passed, the populace and the enemy fighters realized they were not defeated, though U.S. forces mingled among them. Arms, ammunition, and explosives remained abundant. The United States and its coalition partners never generated force-to-population ratios sufficient to control the populations in either Afghanistan or Iraq. The failure of U.S. forces to devastate the opposition, stun the populace into submission, and impose order over the population allowed the conditions for protracted, bloody stability operations to develop.[9] If future presidents and their military advisors are to avoid similar outcomes, they must understand exactly the extent of devastation they must create and the level of control they must impose in order to preclude lengthy and expensive battles with surviving enemy combatants. This consideration should inform discussions about the wisdom of going to war, of the end-state that must be achieved in the prospective war, the campaign plan details, and the expected chain of causality producing victory in any future military operation.

Another lesson from the Second World War is the importance of military governance units. In Germany and Japan, military governance units, backed up by combat forces, were key to consolidating U.S. and allied control and imposing U.S.-allied authority over occupied territory. The Russians, British, Americans, and French all made use of some form of military governance formations to control and administer their zones of occupation in Europe. The United States, Britain, and the Netherlands did likewise with Japan and formerly occupied Japanese territories. As the United States' budget pressures impose force structure reductions on the armed forces, military governance units will probably be unpopular with the military leadership because they claim manpower that might otherwise be devoted to manning combat and combat support units. Force planners, however, should be mindful of the importance of these units in Phase IV operations, how their absence in Iraq and Afghanistan has contributed to the unhappy outcomes that eventuated, and how these outcomes have undermined the use of military force as a viable policy option.

The Unintended Consequences of Rules of Engagement

According to Joint Pub 1-02, the Dictionary of Military and Associated Terms, rules of engagement (ROE) are "directives issued by competent military authority that delineate the circumstances and limitations under which United States forces will initiate and/or continue combat engagement with other forces encountered."[10] What seems so straightforward in principle, however, has generated confusion, frustration, and perhaps more serious consequences on the battlefields of the CENTCOM area of responsibility. Air Force special operations gunships during Operation Iraqi Freedom (OIF) often returned home before sunrise without firing a shot. Sometimes it was because there was nothing to shoot at; more often, it was because the aircrew could not satisfy the engagement criteria specified in the ROE despite having targets in their sights.[11] It has also become commonplace to put military lawyers in joint operations centers (JOCs) and similar command-and-control headquarters, sometimes down to battalion level, to advise commanders on the permissible use of force as combat unfolds.[12] Some troops, including Army Rangers and Navy SEALs, view the ROE as impediments to combat operations and believe that the rules put them and their buddies at greater risk. As one former Ranger put it, "I didn't join to be sacrificed."[13]

The U.S. Army *Operational Law Handbook*[14] notes that ROE distinguish between two functions: the use of force as part of the right to self-defense for individual soldiers and their units, and the use of force for mission accomplishment. According to the Handbook:

> The [ROE] distinguish between the right and obligation of self-defense, and the use of force for the accomplishment of an assigned mission. Authority to use force in mission accomplishment may be limited in light of political, military, or legal concerns, but such limitations have NO impact on a commander's right and obligation of self-defense. Further, although commanders may limit individual self-defense, commanders will always retain the inherent right and obligation to exercise unit self- defense. However, JAs [judge advocates, military lawyers] must be aware that the line between action for mission accomplishment and action in self-defense is not always clear. Distinctions between mission, accomplishment, and self-defense, and between offensive and defensive operations, may vary based on the level of command, array of forces, and circumstances on the ground.[15]

Despite the clear commitment in the handbook to self-defense, some authorities believe increases in casualties can be linked directly to changes in the ROE. Reporter Rowan Scarborough cites Wayne Simmons, a former intelligence official in Afghanistan, who claims:

> In Afghanistan, the [rules of engagement] that were put in place in 2009 and 2010 have created hesitation and confusion for our war fighters. . . . It is no accident nor a coincidence that from January 2009 to August of 2010, coinciding with the Obama/McChrystal radical change of the [rules of engagement], casualties more than doubled.[16]

Lt. Col. David Bolgiano, U.S. Air Force, author of *Combat Self-Defense: Saving America's Warriors From Risk-Averse Commanders and Their Lawyers*, points out other issues with the rules of engagement. In a PBS *Frontline* interview, he observed:

> The rules of engagement, strictly speaking . . . contain a caveat . . . that nothing in these rules of engagement shall limit the right, inherent right, of self-defense. And what's left unspoken but is legally significant, is the phrase, when confronted with "an imminent threat of death or serious bodily injury." . . . Most of those threats, however, don't require a deadly force response. So the rules and the body of law that underpins those rules all concern discerning what is a reasonable response to

> an imminent threat of death or serious bodily injury. The confusion over the inherent right of self-defense comes from assessing judgment-based shootings after the fact that, in the clear vision of 20/20 hindsight, may not appear to be reasonable when, in fact, by law and by tactics, they were.
>
> And that's where a lot of the confusion comes from. It comes from folks that don't understand the tactical dynamics of an encounter and wish to impose, one, either their Hollywood notion of what's reasonable, or two, try to judge the person by what is learned in the clear vision of 20/20 hindsight.[17]

Another consideration is the relationship between the ROE and the chain of causality that is driving military operations. Ideally, the ROE should support the chain of causality and its battlefield effects. Where the rules of engagement constrain the use of military force, the benefits of the constraints should be obvious when viewed in light of the chain of causality.

Potential Problems and Consequences

No one (certainly not this author) wants to see innocent people injured or killed simply because they live in a country at war. Of course considerations for their safety and welfare should influence the conduct of military operations. Nevertheless, it seems that the rules of engagement have become polluted with the "hearts and minds" notions from counter-insurgency orthodoxy, which in turn has led to policies that undermine military effectiveness by adopting flawed chains of causality (e.g., that winning hearts and minds of the locals will result in tips and intelligence that help defeat the enemy). Despite tips and intelligence following the awakening in Iraq, the enemy persists, making the hearts and minds link in the chain of causality a dubious one. The consequences could manifest themselves in the form of several potentially serious problems.

First, there might be an increase in U.S. casualties, accompanied by a decline in the morale of U.S. forces, which could embolden the enemy. A relationship between more restrictive rules of engagement and greater numbers of casualties appears to exist in Afghanistan, as Wayne Simmons claimed. Small, all-volunteer forces may be especially sensitive to casualties, even if the republic they protect is not. ROE that, with the president's approval, pursue a flawed chain of causality that is not producing victory but exposes its volunteer force to increased danger is a large problem the uniformed military cannot address on its own. It will need congressional support and may be getting it. In January 2016, Representative Ryan Zinke (R-Montana) called for an investigation into the death of Special Forces Staff Sergeant Matthew McClintock, who was killed in action in Afghanistan in circumstances that suggested the ROE may have prevented McClintock and his comrades from receiving fire support from an available AC-130 gunship, and delayed the dispatch of quick reaction forces to aid them.[18]

Second, the prevailing ROE reduce operational effectiveness and decisiveness. Effectiveness suffers because operations driven by a bad chain of causality are less efficient than those driven by a sound chain of causality. Decisiveness suffers because battlefield success takes longer than it otherwise might. Congress should bring these observations to the president's attention and demand that he or she address them.

Third, because the enemy is under less pressure as a result of the ROE restrictions, it enjoys greater freedom of action to exert its will on the local population—a perverse twist of events, given that the ROE were influenced by a desire to build favor and support among the populace. Decent force-to-population and force-to-space ratios could be generated in some instances to reduce the enemy's freedom of action despite the ROE restrictions if the administration appreciates the necessity of doing so. Congress could provide that appreciation.

Actions Congress Should Consider

In any future conversations with the White House contemplating an Authorization to Use Military Force, Congress should insist that the president and secretary of defense demonstrate that the size of the force package under consideration is sufficient to compensate for or provide offsets against constraints imposed by the ROE. For example, if the enemy is commingled among an otherwise neutral population, the administration should be required to affirm that the forces deployed will be large enough to do cordon-and-search and similar population-oriented operations to separate the enemy from the populace.

In addition, Congress should require the Department of Defense to undertake a review of recent ROE and their overall effects on the conduct of operations, destruction of the enemy, and the reaction of the local population from 2001 to the present. The objective of the study should be to evaluate the various elements in the ROE and their overall contributions to operational success—assuming that success is the outcome.

Moreover, Congress should support combatant commanders by insisting that the president, as commander-in-chief, adopt a more aggressive policy to identify Declared Hostile Forces within their theaters. Declared Hostile Forces, according to the *Operational Law Handbook*, can include:

> Any civilian, paramilitary, or military force, or terrorist that has been declared hostile by appropriate U.S. authority. Once a force is declared to be "hostile," U.S. units may engage it without observing a hostile act or demonstration of hostile intent; i.e., the basis for engagement shifts from conduct to status.[19]

The objective of the effort should be to prevent the maximum number of actors in a theater of war from enjoying the protections offered by the ROE by designating them Declared Hostile Forces.

Insights from Analysis of Congressional Research Service Reports on U.S. Military Deployments

The Congressional Research Service (CRS) published a report summarizing the instances of use of U.S. military force abroad for the years 1798 through 2014.[20] Research for this chapter considered the entries for the years 1950 through 2014 to capture the deployments of the Cold War and current eras. The reasons for deploying U.S. forces can be organized into six categories:

- Instances of deterrence and shows of force
- Stability, security, peace support, and advisory operations
- Limited objective combat operations
- Noncombatant Evacuation Operations (NEOs)
- Humanitarian Assistance and Disaster Relief (HADR) operations
- Regime change

"Instances of deterrence" accounts for U.S. forces deployed in Germany as part of the North Atlantic Treaty Organization deterrent posture. Shows of force include demonstrations of resolve such as the Berlin Airlift and armed confrontations such as those at Berlin's famous Checkpoint Charlie. There were forty instances of deterrence and shows of force between 1950 and 2013.[21]

Stability, security, peace support, and advisory operations include activities by the Joint Special Operations Task Force Horn of Africa (JSOTF-HOA) that sought to improve stability in that region of the world by working with local friendly forces to prevent Islamist incursions. There were forty-two such entries in the CRS report.[22]

Operation Allied Force in Kosovo during 1999 exemplifies limited objective combat operations. The Serbian government remained intact, but was forced to abandon its policies of murdering its citizens of Albanian ethnicity. The CRS report contains twenty-nine instances between 1950 and 2013.[23]

Noncombatant Evacuation Operations (NEOs) remove U.S. and friendly foreign nationals from areas where security is deteriorating. Embassy evacuations in strife-ridden countries may be the most frequent example. There were twenty-four NEOs during the years in question.

Humanitarian Assistance and Disaster Relief (HADR) operations are military interventions to deliver humanitarian assistance and disaster relief under circumstances too dangerous for international relief organizations to operate on their own. U.S. efforts in Somalia, described earlier in this book, offer a typical example. Fourteen HADRs took place between 1950 and 2013.

"Regime change" involves the removal of a country's government in order to accomplish U.S. objectives. There are four cases in the period 1950 to 2014: U.S. support for the coup that removed the Diem brothers from power in South Vietnam in 1963, the U.S. forcible removal of General Manuel Noriega from Panama, and Operation Enduring Freedom (OEF) and OIF that felled the Taliban and Saddam Hussein governments in Kabul and Baghdad, respectively.

Three of Four Regime Change Operations Created Tragedies

Its complicity with the conspirators who overthrew the Diem brothers in Saigon bound the United States to the successor strongmen. Along the way, Washington lost whatever influence it had on South Vietnam's domestic policy, a realm that was arguably critical to neutralizing the Viet Cong, the indigenous communist militia. The South Vietnamese leaders effectively harnessed the strategic might of the United States for their own purposes, although neither they nor their U.S. advisors developed plans that could have actually halted the war on favorable terms for Saigon. The result was a long, bloody war that Saigon could not win, but one that it would surely lose once the U.S. Congress shut down financial and material support. Thousands of Vietnamese who cast their lot with the United States suffered the consequences.

Afghanistan and Iraq failed in large part because the United States and its partners could not manage the stability operations that should have led to peaceful transfers of authority to locally selected governments. That is, the United States and its partners could not find and kill the numbers of enemy combatants necessary to deter foreign fighters from flooding into these countries, creating a window of opportunity in which newly ensconced governments had the time to find their footing. Nor could the coalition forces exert control over the population and countryside necessary to constrain enemy movements and communications. The handovers to local authorities proceeded, but with fierce resistance still loose in the countryside and very capable of challenging the fragile governments for power. Of course, there was more to it than that. In Iraq, President Maliki's systematic isolation of the Sunni minority undermined his government's legitimacy and sapped the security force's will to confront ISIS and other challengers. In Afghanistan, the United States never fully grasped the role Pakistan was playing[24] and therefore never held Islamabad responsible for its actions that undermined progress, stability, and eventual peace. Secretary of Defense Rumsfeld identified the problem—but failed to address it—early on: "How to fight an enemy present in numerous countries with which we were not at war?"[25] Future administrations must be careful to analyze their war plans in detail, identify all of the key actors, and craft plans to deter, compel, and coerce them as necessary in order for U.S. forces to be successful.[26] In Pakistan's case, appropriate actions might have been to engage China as an agent of influence, given the special relationship between Beijing and Islamabad, as well as China's growing apprehension about unrest and instability in Pakistan prompting similar instabilities on Chinese territory. Washington might also have encouraged India toward development assistance in Afghanistan, an action that would have benefited Afghan society and its economy while raising concerns in Islamabad about Pakistan being outflanked by India. India might have viewed such encouragement favorably if China continued to stymy India's UN-based diplomatic efforts to address Pakistan's terrorist connections.[27]

Insurgency and Irregular Warfare Remain Widespread Problems

U.S. military deployments over the years in question served many purposes, from deterrence and shows of force to regime change and more humanitarian goals. These deployments were often the substance of the containment strategy the West pursued against the Soviet Union and its subject states. In some instances, U.S. forces were responding to the negative effects of de-colonization and the civil wars that resulted (and Marxist-Leninist efforts to exploit the strife to expand their influence). In many instances, the enemy on the ground was practicing insurgency or some form of irregular warfare: the very forms of warfare that confound U.S. armed forces to this day. Therefore, if the United States were to create capabilities for fighting insurgencies and irregular adversaries like those suggested in chapter 4 of this book, it would be making investments that could improve future U.S. military performance considerably. And, frankly, if the United States cannot master counter-insurgency and stability operations to overcome the world's irregular warriors, it constitutes a major Achilles' heel and significantly undermines the use of military force as a policy option.

Limited Objective Combat Operations Did Not Produce Reliable Outcomes

Of the twenty-nine instances that took place during the years in question, only six might qualify as victories. These include Kosovo, Bosnia, the Philippines, Grenada, the Dominican Republic, and South Korea: one in five, approximately. Kosovo and South Korea remain frozen conflicts that require expensive military coalitions to maintain the status quo. Bosnia's outcome produced a Serb rump state that continues to challenge peace and stability. The Philippines and the Dominican Republic may be the most successful and carry the least complications. Thus, simply shying away from regime change and seeking to treat U.S. security challenges through the mechanism of limited objective operations will not be a wholly satisfactory solution; the United States still needs careful analysis of the military problem to be treated, development of a clear chain of causality for military actions, and a campaign plan that is appropriately destructive if it is to prevail. In other words, the U.S. military and its civilian leadership must invest the time—even in crisis circumstances—to understand the threat confronting them, craft an appropriate response, and execute the plan with requisite thoroughness and attention to the Phase IV endgame.

Success Breeds Success and Vice Versa

Measuring the frequency of U.S. military interventions is revealing. Between 1950 and 1991, the average number of interventions per year was 1.5. The years from 1992 through 2013 averaged 5.6, and this average may be low because some of the later entries do not report actual numbers but simply state that "several countries in the region" may have witnessed U.S. military actions.

What accounts for this increase in the frequency of U.S. military interventions? One explanation would attribute the lower, pre-1992 figure to the stability imposed by the Cold War, and the higher number of deployments annually thereafter to the rise of Islamic terrorism. But, according to the Stockholm International Peace Research Institute 2015 Yearbook, despite the outbreak of violence between Russia and Ukraine and civil wars in the Middle East, "the world remains more peaceful today than in all periods before the year 2000."[28] It is, therefore, unlikely that the frequency of U.S. military deployments is the product of increased violence around the world. An alternative explanation is *The Atlantic* contributor James Fallows's "Chickenhawk Nation" thesis, which argues that America's lack of connection to its armed forces makes it easy to deploy them into harm's way. The lack of familiarity with the military also works to reduce serious debate about the wisdom of specific deployments.[29] According to Fallows, these circumstances have led to U.S. forces being deployed again and again "to fight unwinnable wars."

An antidote for such practices could be the robust civil-military relationship based upon congressionally prescribed processes described in chapter 2 of this book. That is, if the population is disinterested, the senior civilians lacking in experience, and the senior uniformed officers not prepared to speak candidly to civilian authorities or have no advice to offer,[30] then a process that lays out specific roles and expectations of advice from all parties might help an otherwise uncertain group at least identify the key issues they must understand and help them find clues as to where to look for useful information.

Alternative Approaches to Afghanistan and Iraq

No book of this sort, criticizing presidential administrations and the Pentagon for their mishandling of America's recent wars, can come to the end making observations and recommendations, as the next chapter does, without offering credible alternative approaches to the wars in Afghanistan and Iraq. The attempt that follows employs a number of important assumptions: first, that the Bush administration's stated concerns about Afghanistan and Iraq were sincere and provided the actual motivation and rationale for those operations; second, that if mission analysis from the professional military indicated a need for important force structure changes (e.g., growing a significant capability to find, secure, and demilitarize weapons of mass destruction [WMD] under field conditions) in order for the operation to be successful, a priority effort would attempt to build the required forces, and train and equip them for these roles. This last assumption is an optimistic one in light of author Brian Michael Linn's analysis, which documents the army's tendency to prepare for wars that seldom occur and to ignore those it must actually fight—a tendency that could be prevented by sound civil-military relations.[31]

Afghanistan

According to the 9/11 Commission Report,[32] the Bush administration intended to punish the perpetrators of the September 11, 2001, attacks on the United States and their supporters. Afghanistan, with its Taliban government and al Qaeda guests, was a first step in a broader strategy to eliminate terrorism. Pakistan's critical role in shaping the outcome in Afghanistan was identified at the very beginning.[33] The United States demanded Pakistan's cooperation to:

- stop al Qaeda operatives at the border and end all logistical support to bin Laden;
- give the United States blanket overflight and landing rights for all necessary military and intelligence operations;
- provide territorial access to U.S. and allied military intelligence and other personnel to conduct operations against al Qaeda;
- provide the United States with intelligence information;

- cut off all shipments of fuel to the Taliban and stop recruits from going to Afghanistan; and
- if the evidence implicated bin Laden and al Qaeda and the Taliban continued to harbor them, to break relations with the Taliban government.[34]

Treating Pakistan as an ally was the administration's first mistake. An alternative analysis of the military problems confronting the United States in Afghanistan, informed by the president's guidance, identifies three main tasks: first, drive the Taliban government from power for its complicity with al Qaeda; second, destroy al Qaeda in retaliation for its attacks on the United States; third, deter Pakistan from continuing its support for Taliban and al Qaeda—wherever and in whatever form it takes place.

The alternative military plan to remove the Taliban and destroy al Qaeda would not differ significantly from what in fact took place. A key mistake in the actual plan was that it depended too heavily on indigenous forces of the Northern Alliance for success and deployed too few U.S. forces that could act decisively if the local allies proved slow to respond. The main difference in an alternative approach would be a larger overall U.S. force package so that local commanders had adequate, reliable forces at their disposal. Dependence on the Northern Alliance and the constraints imposed by the imperatives to work with local militias would have been minimized if more U.S. forces had been available. A larger force package would have allowed for U.S. forces to be deployed in blocking positions to prevent the enemy's flight from Tora Bora across the frontier into Pakistan: according to Peter Bergen, writing in *The New Republic*, "one of the greatest blunders in recent U.S. history."[35] In this alternative campaign, al Qaeda would have been vanquished at Tora Bora by U.S. forces.

Clear notice to Pakistan that failure to stop enemy fighters entering its territory would produce severe reprisals would have been an important feature of the alternative military plan. If evidence developed indicating that Pakistan had allowed Taliban/al Qaeda personnel to escape into its territory, the plan would have called for punitive strikes against selected Pakistani military facilities. The alternative military plan would also signal India that the United States supports its counter-terrorism efforts with respect to Pakistan, offering New Delhi a green light for more aggressive action against Islamabad.

Phase IV and Phase V (transfer of authority) operations would be very limited, enduring only long enough to see a successor government ensconced in power. Once the Karzai interim government was installed and the Afghan people began charting their own future free from the specter of al Qaeda, the chapter of U.S. military involvement would come to a close, and U.S. forces would withdraw. There would be no nation building, because nation building is expensive and ineffective. Figure 5.1 illustrates this point.

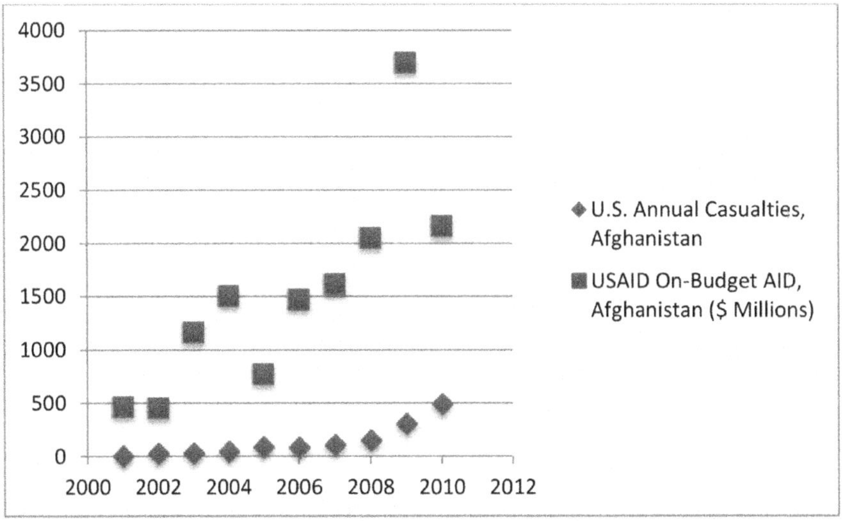

Figure 5.1. U.S. Agency for International Development On-Budget Assistance versus U.S. Casualties. *Source*: U.S. Mission Afghanistan, U.S. Foreign Assistance for Afghanistan Post Performance Management Plan 2011–2012, Volume 1 Summary (p. 122).

An implicit hypothesis about nation building is that as the quality of life improves for local people as a result of nation-building projects, the people will come to see U.S. and allied forces in a positive light, and cooperate with these forces when the risks of doing so are minimal (e.g., sharing information about the enemy). But, as the figure shows, aid grew steadily, but so did U.S. casualties. The implicit hypothesis is clearly false. Several alternative hypotheses might explain the relationship between nation building and casualties. One such hypothesis is that casualties resulted from military operations meant to secure nation-building projects. That is, al Qaeda saw nation-building projects as attractive targets where they could kill U.S. forces. This hypothesis might be confirmed if we could establish that U.S. casualties occurred near nation-building projects (unfortunately, we cannot establish such a relationship). Another hypothesis claims that U.S. casualties and U.S. Agency for International Development spending were unrelated. Evidence in support of this hypothesis would be that many al Qaeda fighters were foreigners who cared little about the living conditions among ordinary Afghan people and pursued their efforts to kill U.S. and coalition forces no matter how relations between the populace and the U.S.-led coalition evolved. An alternative interpretation of these statistics is that they are unrelated because ordinary Afghans were largely unmoved by the nation-building efforts and too terrorized by the local Taliban/al Qaeda combatants to offer the U.S.

forces meaningful information and tips that might have prevented the casualties in question (although denser U.S. force-to-population ratios might have reduced enemy influence and terror with respect to the population). Either way, nation building appears both expensive and ineffective.

The plan for deterring Pakistan would be based on the realization that Pakistan is a hostile state. Washington would be explicit with Islamabad about its expectations for Pakistan's behavior regarding Afghanistan: that Taliban and al Qaeda personnel found on Pakistan's territory will be turned over to the United States, that Pakistan and Afghanistan will cooperate to defend their mutual border and prevent illegal crossings, and that Pakistan will cease all forms of support to the Taliban and al Qaeda or face a combination of sanctions and military action.[36]

With Pakistan warned off future interference and Afghan officials enjoying a reasonable opportunity to establish sound governance and rapport with the population before its enemies can reorganize and present a renewed challenge, the U.S. and its coalition partners would withdraw. Doing so would reduce the incentives for Afghan factional infighting, made possible by the confidence of knowing that the United States and its partners were around if things deteriorated. Instead, without the United States and the coalition, Afghan officials would have incentives to cooperate for their own survival and forge a robust government and counter-terrorism plan to prevent the resurgence of the Taliban and al Qaeda.

Of course, things could go wrong. Pakistan might not be persuaded by U.S. threats and promises and require sanctions or punitive attacks to force it to abandon its policy of support for the Taliban and al Qaeda. The Taliban and al Qaeda might manage a resurgence that again imperils Afghanistan's security. The U.S. administration would have to take appropriate action, including, perhaps, a new deployment of forces, but operations in Afghanistan were just the first steps in what was always envisioned as a longer war against terrorism. Returning to engage again would not, in that context, be indicative of failure, but rather a continuing need to treat the problem. However, if a return engagement could be avoided, the savings in terms of lives lost and treasure spent could be significant. The United States would save at least $30 billion that it spent on Pakistan since 2001.[37] It would save another $14.5 billion on U.S. Agency for International Development nation-building assistance to Afghanistan if aid stopped after 2002. A total of 2,206 U.S. lives would also be spared if U.S. military operations stopped at the end of 2002.

Iraq

The Bush administration considered action against Iraq in the days immediately following 9/11.[38] The president decided to defer action until later, and by

2003, conditions seemed right. The Taliban had been routed from Afghanistan, al Qaeda was on its heels, and the situation was deteriorating in Iraq. In addition to its concerns about the growing risks associated with maintenance of the no-fly zones over northern and southern Iraq, Saddam's apparent success at skirting UN-imposed sanctions, and suspicions of Iraqi support for terrorism, the administration renewed its concerns about Iraqi weapons of mass destruction. These concerns were amplified by Iraq's suspicious behavior. Iraq was a signatory of the Nonproliferation Treaty and was subject to International Atomic Energy Agency safeguards for its nuclear energy infrastructure and materials. However, Baghdad was failing to cooperate with the inspectors; moreover, it was clear that the regime was intimidating its nuclear scientists and technicians, holding their families hostage. Together with intelligence reports of WMD, these circumstances strengthened the United States' assertion that Iraq was not in compliance with UN Resolution 1441, which required Baghdad to adhere to the cease-fire agreements it had signed at the end of the First Gulf War in 1991. These obligations included prohibitions against WMD and other weapons systems, including missiles.[39] In addition to the concerns that were specific to Iraq, the Bush administration was also concerned about broader, regional stability. During the administration's deliberations, Secretary of State Powell warned that action in Iraq could destabilize the entire region.[40]

In an alternative analysis of the challenges the United States and its partners faced in Iraq, four distinct military problems emerge. The first is the question of how to remove the regime without destabilizing the entire region. The second is how to attain freedom of action in the country that will allow searching for, finding, securing, and demilitarizing WMD under field conditions. The third military problem is how to identify actors within Iraq who supported terrorist organizations and punish them. The final military problem is how to maintain order and stability in the country until authority can be transferred to a successor government. The sub-sections that follow treat these issues.

Regime Change and Regional Stability

The key to regime change without destabilizing the whole region lies in the quality of the U.S. victory. The Iraqi armed forces would have to be decisively defeated. Those elements of the Iraqi forces likely to be irredeemably hostile to the United States would have to be destroyed: the Republican Guard and the Fedayeen Saddam (Ba'ath Party paramilitary formation). The remainder of the Iraqi forces would have to be destroyed to the point that the survivors surrendered on the battlefield and are taken into U.S.-coalition custody. This detail is crucial. Under such circumstances, the survivors would know they had been defeated and that further resistance would be futile.

The remnants of Saddam's forces would be rounded up and placed in prisoner of war (POW) camps. Their treatment would be humane and monitored by the United Nations and Red Crescent or similar organizations. Public affairs announcements would be clear that their detention is temporary and that everyone in custody will be released when the United States finishes its mission in Iraq, removing the country's WMD and the capability to manufacture them. The public affairs officials would be explicit about the linkage between return of the POWs and the United States' military presence: the sooner U.S. forces are convinced they have found and destroyed the last vestiges of Saddam's WMD program, the sooner the POWs will be returned to their loved ones.

The United States would increase its military presence after the cease-fire to ensure it can control the population, infrastructure, the border, and the highway system. U.S. troop deployments would be "for the duration" rather than rotational, and would require the full mobilization of the reserve components including military governance units. Such a policy could not be sustained indefinitely, but it would provide enough U.S. forces to exert positive control over key aspects of life in Iraq, prevent looting and destruction of public property, and operate an effective military government.

Attaining Freedom of Action in Iraq

Freedom of action means the ability to move around the country without indigenous interference, enjoying full access to Iraq's facilities and institutions, especially those associated with the country's WMD efforts. Attaining freedom of action requires overwhelming military force to convince the population that interference is futile and its consequences dire. An important element of attaining freedom of action involves disarming and controlling the population. Population control would be based in part on the methods of instrumentation and tracking described in chapter 4 and in part on an extensive and aggressive U.S. military posture enforcing curfews, running checkpoints, and otherwise imposing its will on the population.

U.S. and coalition officials engaged in the hunt for WMD would call upon Iraqi citizens for assistance and information. The military government would also offer generous rewards for information leading to the discovery and seizure of WMD and other dangerous weapons and materials (e.g., unexploded ordnance, abandoned artillery ammunition). Military intelligence would establish networks of informants so that citizens could provide information and receive payment for it without their neighbors becoming aware of their cooperation.

The U.S. military would need an appropriate organization to search for WMD and deal with them. In the case of OIF, it had one, based on the 20th Support Command (now 20th CBRNE Command).[41] The unit's operations were not the top priority, however, and it was not large enough to deploy detachments promptly to known and suspected sites, much less conduct a methodical search of all Iraqi infrastructure. Nor did the 20th Support Command have the necessary combat capabilities to force its way into a facility of interest and secure the site until it completed its operations there. In this alternative approach to OIF, the U.S. Army would have deployed several such units, each paired with combat forces that could guarantee them access to whatever site they were investigating and secure it until counter-WMD operations there were concluded. And the search for WMD would have very high priority in the overall operation.

Identifying Supporters of Terrorism

OIF effectively treated identification and arrest of supporters of terrorism as a lesser-included case in allied efforts to find the key members of the Saddam regime. The maximum number of Iraqis in detention at Guantanamo Bay was sixteen: perhaps a figure indicative of U.S. success in identifying supporters of terrorism.[42] In the alternative approach, having bought the cooperation of former regime personnel, the military government would take a two-pronged approach to the task. First, U.S. officials would scrutinize the former regime's documents to find the paper trail tying Saddam's regime to terrorism. The investigation would include a review of the files on all intelligence personnel, military and otherwise. Second, the military government would enlist the population's help by operating a tip line and offering rewards for information that leads to convictions for supporting terrorism. The investigation would include examination of influential mullahs and contributions from their congregations that may have found their way into terrorist accounts. The United States would also ask for assistance from the Kurds and other minorities who may have their own insights into how the Saddam regime supported terrorism.

Maintaining Order and Stability

Iraq is a big country with twenty-six million people. Maintaining order and stability is a critical task from both a U.S. and an Iraqi perspective. It is important to Washington because it expedites the search for WMD and related weapons systems; because an orderly, stable Iraq accelerates the transition to Iraq authorities; and because an orderly, stable Iraq requires fewer U.S. and coalition forces to ensure order and stability. From the Iraqi perspective, order and stability mean a return to normalcy and hasten the departure of infidels from their territory. To attain these conditions, the U.S. and its partners will deal with the cities and their populations as described earlier in this book and install military governance units supported by combat forces. The cities will thus be inhospitable to militias and small bands of Saddam faithful. Those who want to contest the authority of the military government will have to go elsewhere to organize, train, and plot. Outside the cities and towns, they can be located, identified, tracked, and neutralized by coalition forces. In the cities, special operations forces (SOF), the military police, military intelligence, and other U.S. formations can rout them out.

The population, on life support to the military government for food, fuel, electricity—virtually everything—will be grudgingly compliant. Towns and neighborhoods that cooperate and provide useful information to the authorities that helps prevent acts of violence can be rewarded with larger rations, access to telecommunications, and similar inducements to further cooperation.

The Alternative Campaign Plan

To treat the military problems just described, the United States and its partners will create and execute a large-scale campaign. As in the actual OIF, the alternative campaign opens with air power, but on a larger, more sustained basis to ensure destruction of its targets. The targets begin with Iraqi air defenses, which must be destroyed to allow airmen freedom of action in the skies. WMD delivery systems—large-caliber field artillery units, rocket and missile units, and the Iraqi air force—would be in the crosshairs for cruise missile and drone attacks from the earliest moments of the war. Far better to destroy these in garrison than to hunt them down throughout the whole country. Regime leadership targets (e.g., Ba'ath Party headquarters, Saddam's multiple palaces) would be next.

Once the skies are safe for allied air forces, bombers and attack aircraft would begin attacks on the Republican Guard and subsequently, after the Guard's annihilation, the rest of the armed forces. After the military and regime targets, air power would turn to infrastructure (e.g., water pipelines, electricity, telecommunications) and cities and towns.

Many readers will balk at the notion of bombing cities, but doing so is vitally important to the ultimate success of the war.[43] The resulting destruction and death will destroy hard-core resistance and convince the remaining population that it cannot resist and that survival depends upon getting a cease-fire and cooperating with the victors—a message that would accompany the bombing in the form of leaflets and radio broadcasts. The bombing also destroys food stocks and other necessities so that the United States and its partners can ration these commodities to the populace as an instrument of control. The ultimate result might be psychological overmatch—the condition resulting in Iraqi minds from swift, sure, thorough destruction and the realization that their only chance to outlive the current hell is to submit to coalition authority.

When the ground war begins, the CBRNE units and their combat arms escort units will head directly for known and suspected WMD sites, seize them from any lingering security forces, and begin their assessment, render-safe, and demilitarization of WMD and related materials. Simultaneously, the major combat units will attack toward their assigned objectives, typically towns and cities, destroying any remaining resistance and accepting Iraqi surrenders along the way. Right behind them, military governance units will move into the captured towns, establishing order, passing out transistor radios so they can listen to instructions about how to behave upon the arrival of coalition forces, offering news about the new military government, and providing broadcasts from prominent co-opted Iraqis, urging cooperation. These military governance units will offer medical care, provide food and water, and begin the laborious process of issuing RFID identity cards and ration books and deploying the system of social controls.

As the coalition force becomes settled in Baghdad, coalition authorities will invite the United Nations to help restore order and services, and to advise on the transfer of authority from the military government to an Iraqi successor government. The United Nations would also organize a donors' conference to assist Iraq in recovering from the war. Arabic broadcasts will reach surviving Iraqi military personnel, promising safe passage and humane treatment upon their surrender. Coalition aircraft will scatter safe conduct passes over areas thought to hold Iraqi armed forces survivors and stragglers, reinforcing the radio broadcast messages.

Coalition forces would flood the country, performing stability operations to safeguard the country's frontiers and secure its population and infrastructure. SOF would remove opposition figures that might contest the authority of the coalition and mobilize militias against it (e.g., Muqtada al-Sadr, who mobilized his followers to create an Islamist government and resist U.S. influence). The military government would transfer authority to the Iraqis or to the United Nations (in the event that the Iraqis cannot reach consensus on an interim government) as soon as senior U.S. officials are satisfied that the WMD threat has been eliminated. At that point, the U.S. and coalition forces begin their withdrawal.

Unlike the Hobbesian reality that prevailed from 2004 onward in varying degrees, the post–cease-fire Iraq in the alternative case would be relatively safe and stable. The population, subject to the coalition's social controls, would have little means to resist, deploy improvised explosion devices, or carry out raids against coalition facilities. They would have little freedom of movement, limited ability to communicate beyond their immediate neighbors, and few means to make war. The CBRNE units would have denuded the landscape of unexploded ordnance. Cash-for-guns programs would police up many of the small arms lying about, and the draw of larger food and fuel rations in return for cooperation would be significant. Moreover, the public face of the military government would be stern and subtly menacing, constantly reinforcing the message that acts of violence against the U.S.-led coalition will be dealt with aggressively and that the coalition's presence is temporary, that Iraqi cooperation will speed its departure. With some luck and credible information operations, Iraqis would recognize the opportunity to create new government institutions. The United States would be indifferent to the details of the internal Iraqi settlement, as long as it seemed likely to produce stability. Either way, the United States would withdraw as soon as it was certain the Iraqi WMD program had been effectively destroyed.

Conclusions and Observations

As it has always been, waging war is uncertain business, and limiting the expansiveness of war's objectives does little to improve the prospects of success. Three out of four wars that produced regime change ended tragically. Limited objective combat operations, as the CRS data demonstrate, produced mixed outcomes. Counter-insurgency, as the RAND data reflect, has been a long odds proposition for states attempting it, and success depends on factors that are often absent in states struggling against an insurgency. Nevertheless, when all other policy options have failed and the United States must defend its interests through the use of military force, this chapter suggests several key considerations.

Scale and duration of violence is critical to destroy the enemy's will and ability to fight, cause the population to prefer capitulation to continuation of the violence, and to compel the regime to accept U.S. terms (e.g., abandon actions and policies that threaten U.S. interests). In instances where the enemy population is deemed hostile to the United States and actively supports its regime's war-making efforts, the level of destruction required to make it capitulate may be vast. When the threats to U.S. vital interests warrant it, the president and senior defense officials must avoid half measures, use the force required, and be prepared to defend the extent of the violence to world public opinion, and perhaps especially, to allies and coalition partners. If the administration is unwilling to defend such an aggressive and destructive campaign, it must reconsider its objectives in going to war and establish outcomes that seem realistic with a more constrained use of military force.

Where seizure and occupation of enemy territory are part of the plan, U.S. forces must be prepared to act as the local authorities until arrangements can be made to return administration to a local government or to international organizations. Re-creation of military governance units that are organized, trained, and equipped for this role is a historically sound approach for stability operations pending transfer of authority. If they are to be successful, military governance units must be scaled to the size of the population: roughly 309 personnel for every ten thousand population served.[44] The military governance units must be efficient enough and effective enough so that the population prefers their services to those of a government in exile, local militia, or similar competitor for influence. The U.S. military should be directed to develop doctrine and dedicated force structure for these Phase IV tasks.

In the absence of manpower and local knowledge that are essential to traditional counter-insurgency solutions, the United States should develop technologies for population and territorial control approaching Orwellian levels of dominance. Doing so will support restoration of social order, improve local security for both the population and U.S. forces, and deprive the enemy of its ability to move about, communicate, and attack the security forces. British and Chinese use of surveillance cameras in their largest cities demonstrates the technical feasibility and the contribution to security that instrumentation can make.

Notes

1. Nicholas Stargardt, *The German War: A Nation Under Arms 1939–1945* (New York: Basic Books, 2015), 535.
2. Ibid., 541.
3. http://warchronicle.com/numbers/WWII/deaths.htm.
4. Ibid.
5. Paul Kecskemeti, *Strategic Surrender: The Politics of Victory and Defeat* (Santa Monica, CA: RAND Corporation, 1957), chapter 5.
6. Giles MacDonogh, *After the Reich: The Brutal History of the Allied Occupation* (New York: Basic Books, 2007). Although there is some debate regarding the number of Germans killed in the aftermath of the Second World War, MacDonogh claims the number is 2.5 million.
7. David Stafford, *End Game, 1945: The Missing Final Chapter of World War II* (New York: Back Bay Books/Little, Brown, 2007).
8. Karl Dönitz, Grossadmiral, *Zehn Jahre und Zwanzig Tage*, translated by R. H. Stevens as *Memoirs: Ten Years and Twenty Days* (Cleveland: World Publishing, 1959), quoted on Wikipedia.
9. John E. Peters, "A Potential Vulnerability to Precision-Strike Warfare?" *Orbis*, Summer 2004. See also James Clad, "Wasting the Golden Hour in America's Iraq Meltdown," *National Interest*, July-August 2013.

10. Joint Publication 1-02, Department of Defense Dictionary of Military and Associated Terms, http://www.dtic.mil/doctrine/new_pubs/jp1_02.pdf.

11. Author's interviews with AFSOC aircrew members, 2010.

12. *Wall Street Journal*, "Civilians in Crosshairs Slow Troops," http://www.wsj.com/articles/SB10001424052748704751304575079741450028512.

13. Truth and Action website, "U.S. Troops Fed Up with 'Rules of Engagement' That Strengthen the Enemy," http://www.truthandaction.org/us-troops-fed-rules-engagement-strenghten-enemy/.

14. U.S. Army, *Operational Law Handbook: International and Operational Law Department* (Charlottesville, VA: Judge Advocate General's Legal Center & School, U.S. Army, 2010), at CLAMO@hqda.army.mil.

15. Ibid., 75.

16. Quoted in Rowan Scarborough, "Shades of Vietnam: Spike in U.S. Troop Deaths Tied to Stricter Rules of Engagement," *Washington Times*, December 5, 2013, http://www.washingtontimes.com/news/2013/dec/5/increase-in-battlefield-deaths-linked-to-new-rules/?page=all. Records show that casualties increased significantly as Simmons claimed. Military fatalities in Afghanistan for 2008 were 295, 521 in 2009, and 711 in 2010, according to http://icasualties.org/oef/.

17. PBS *Frontline*, "The Rules of Engagement," http://www.pbs.org/wgbh/pages/frontline/haditha/themes/roe.html.

18. Michelle Tan, "Hundreds Support Family of Fallen Green Beret; Lawmakers Seek Answers to Deadly Battle," *Military Times*, January 7, 2016, http://www.armytimes.com/story/military/2016/01/07/hundreds-support-family-fallen-green-beret-lawmakers-seek-answers-deadly-battle/78421900/.

19. *Operational Law Handbook*, 75.

20. Barbara Salazar Torreon, "Instances of Use of United States Armed Forces Abroad, 1798–2014," Congressional Research Service report R42738, September 15, 2014.

21. Four in the Pacific, four in the Caribbean/Americas, five in Europe/Balkans/Caucasus, fifteen in Africa, and the remainder in the Middle East.

22. Seventeen took place in Africa, twelve in the Middle East, six in the Americas, five in Europe/Balkans, and two in Southeast Asia.

23. There were eight cases in the Middle East, eight in Africa, five in Southeast Asia/West Asia, six in Europe/Balkans/Caucasus, and two in the Americas.

24. Carlotta Gall, *The Wrong Enemy: America in Afghanistan, 2001–2014* (Boston: Houghton Mifflin Harcourt, 2014), xiv.

25. Donald Rumsfeld, *Known and Unknown: A Memoir* (New York: Sentinel, 2011), 354.

26. A view endorsed by the Council of Foreign Relation's Alyssa Ayres in her Policy Innovation Memorandum 53, "Why the United States Should Work with India to Stabilize Afghanistan," http://www.cfr.org/afghanistan/why-united-states-should-work-india-stabilize-afghanistan/p36414.

27. Ankit Panda, "Why China Snubbed India on a Pakistan-based Terrorist at the UN," *The Diplomat*, June 25, 2015, http://thediplomat.com/2015/06/why-china-snubbed-india-on-a-pakistan-based-terrorist-at-the-un/.

28. http://www.sipri.org/yearbook/2015/downloadable-files/sipri-yearbook-2015-summary-pdf.

29. James Fallows, "The Tragedy of the American Military," *Atlantic*, http://www.theatlantic.com/magazine/archive/2015/01/the-tragedy-of-the-american-military/383516/.

30. Lieutenant General (U.S. Army, Ret.) Daniel P. Bolger, *Why We Lost: A General's Inside Account of the Iraq and Afghanistan Wars* (Boston: Edmond Dolan Books/Houghton Mifflin Harcourt, 2014), 430, 431.

31. Brian Michael Linn, *The Echo of Battle: The Army's Way of War* (Cambridge, MA: Harvard University Press, 2007).

32. *The 9/11 Commission Report: Final Report of the National Commission on Terrorist Attacks Upon the United States* (New York: W. W. Norton, undated), 330–33.

33. Ibid., 331.

34. Ibid.

35. Cited in David Milne, *Worldmaking: The Art and Science of American Diplomacy* (New York: Farrar, Straus and Giroux, 2015), 439.

36. Georgetown Professor C. Christine Fair and her colleague Sumit Ganguly advocate a similar policy toward Pakistan today. See Fair and Ganguly, "An Unworthy Ally," *Foreign Affairs* (September–October 2015), 160–70.

37. Ibid.

38. *9/11 Commission Report*, 336.

39. See http://www.un.org/Depts/unmovic/documents/1441.pdf.

40. Dale R. Herspring, *The Pentagon and the Presidency: Civil-Military Relations from FDR to George W. Bush* (Lawrence: University Press of Kansas, 2005), 398.

41. "CBRNE" stands for chemical, biological, radiological, nuclear, and explosive. See the unit home page: http://www.cbrne.army.mil.

42. See https://en.wikipedia.org/wiki/Iraqi_detainees_at_Guantanamo_Bay.

43. Bolger, *Why We Lost* (109), notes that the coalition in Iraq "self deterred" and did not use its full force. This forbearance was perceived as weakness by the population and in this author's judgment may have emboldened Iraqis to attack U.S. and coalition forces.

44. This is the median density for municipal services in the United States according to governing.com. The highest density of municipal service workers per ten thousand of population is 494, excluding teachers. The minimum density is 188. See www.governing.com/gov-data/public-workforce-salaries/states...html.

6

Implementation

According to the preceding chapters, the United States faces four main problems with its military instrument. First, its civil-military relations are uneven, and recent experience demonstrates that even seasoned, mutually cooperative teams of senior civilians and generals can produce suboptimal outcomes. Second, the military establishment has lost the ability to produce successful campaign plans. It may be telling that the military services often mistakenly label a variety of official documents "campaign plans" when they are actually administrative instructions for various headquarters functions. Third, the United States has developed a mistaken approach to counterinsurgency and irregular warfare. The services may think they are following canonical approaches developed by the British and French, but they are not. Moreover, the approach that has evolved is generally unsuitable for ground forces of today's size, and likely to become less suitable as the army and Marine Corps undergo force reductions in the years ahead. Finally, there is the inattention to the full menu of military solutions other than regime change. And alternatives are not obvious; certainly limited objective operations are not clearly superior to regime change operations, and the experiences of the Second World War suggest that devastating levels of destruction are needed to overcome an enemy state's resistance and compel it to capitulate.

The president, in his or her role as commander-in-chief, influences each of these problems profoundly. His or her decision-making and executive style are critical to civil-military relations, as are his or her expectations for the competence and abilities of his or her senior defense advisors. The president's ability to explain what he or she wants from options for the use of military force and his or her advisors' ability to understand requests as he or she intends are vital but easily blown off course, as chapter 2 described. Even when the president has made him- or herself clear, things can still go wrong if the president's national security team believes it cannot remind him or her of warfighting objectives that have not yet been accomplished. The commander-in-chief is also central to sound campaign planning. If the president delegates too much to the field commanders (as George H. W. Bush did in the First Gulf War) or refuses to be drawn in to the details of breaking crises and contingency planning (as William J. Clinton did), things can go badly wrong. Alternatively, if the president emphasizes domestic political imperatives over the political-military imperatives of ongoing operations (as Barack Obama has done), fragile and tentative progress can quickly degenerate into a rout. Perhaps most important, the commander-in-chief must satisfy him- or herself that the military has planned adequately for all phases of operations, including the crucial Phase IV operations, which U.S. forces have fumbled from Panama to Afghanistan to Iraq.

Because the commander-in-chief is the president, who is not obligated to act on recommendations from the national security team, and because the Department of Defense is generally incapable of reforms on the scale this book has advocated, this chapter speaks to Congress. The Congress, with its legislative and budgetary authority, is the only remaining federal institution that might address these problems. It can pass legislation specifically to treat these issues, as the individual chapters of this book have suggested. It can embed requirements in the National Defense Authorization Act for the DoD to create new offices and functions such as an Office of Civil-Military Relations, and to conduct studies of issues like the rules of engagement (ROE) question. The Congress has passed the War Powers Resolution requiring presidents to notify Congress on the use of military force. It could similarly pass a resolution requiring presidents to adopt a structured process for civil-military deliberations. Of course, congressional action takes time and may render imperfect solutions. Yet no other federal actor seems able to wield useful power and authority to treat these problems.

This chapter answers the questions, "If one wanted to implement the recommendations from earlier chapters, what would be required?" and "What would it cost?"

Fixing Civil-Military Relations

If Congress agrees with the thrust of the analysis presented in chapter 2—that uneven civil-military relations have been a significant contributor to the

U.S. military's performance shortfalls, and that imposing structure on civil-military consultations would improve decision-making (e.g., by avoiding the pitfalls of System 1 thinking and providing a process for auditing military progress)—it should pass an act similar in type to the Goldwater-Nichols Act of 1986. The new act should require the president, the secretary of defense, the chairman of the Joint Chiefs of Staff, the combatant commanders, and the service chiefs to develop a consultative process to structure their dialogues and periodically audit progress on the planning for and use of military force. The dialogue would, among other topics as the president prefers, include the objectives of the expedition, the specific military problems to be solved, the chain of causality the parties expect will unfold with their successive operations, the end state that will eventuate, the numbers and types of forces required, and, in particular, how these forces will be employed in Phase IV to ensure a satisfactory outcome with no unfinished business. The act should require two discrete processes, including one to guide routine deliberations and another to guide crisis responses.

The act should direct the Office of the Secretary of Defense (OSD) to create an office of civil-military relations to be the institutional home of the consultative process. The office would assist each newly inaugurated president in establishing his/her own approach to conducting the dialogue with the president's military subordinates, document it, and teach it to the other senior participants and their staffs. The office should be headed by a career member of the Civil Service to ensure impartiality and to preclude a presidentially appointed official from short-circuiting the process for expediency's sake during a crisis. The act should direct the office to acquire appropriate analytical support capabilities from the intelligence community and the military services to support course-of-action analysis and decision-making. Just as the executive branch of government does with other critical programs, the Office of Civil-Military Relations should be required to conduct periodic exercises using members from the National Security Council staff, joint staff, and service staffs as role-players for the president, secretary of defense, and others to establish proficiency, discover shortcomings and omissions in the process, and to assist the president and other principals with the process during actual deliberations.

If we imagine the Office of Civil-Military Relations as half the size of OSD's Cost Assessment and Program Evaluation (CAPE) office, it would comprise approximately seventy people and require an annual operating budget of $12.5 million.[1] As with CAPE, the new civil-military relations office might be staffed by a mix of civil servants and uniformed military personnel. Seventy-some people and $12.5 million could be a very smart investment if the result is better outcomes from the civil-military relationship and the deliberations among senior officials that drive future military operations.

Restoring Campaign Planning Proficiency

If Congress is prepared to get serious about restoring the armed forces' ability to do campaign planning, one of the first steps it could take would be to direct a change to the way it is taught. Congress could usefully direct a major effort at curriculum reform that would set aside cases from the Civil War in favor of more modern instances of campaigning—including foreign examples—where the commander-in-chief faced political-military constraints but nevertheless prevailed and why, and where the military problems to be addressed were buried beneath social, ethnic, and economic disputes. The point would be to call upon recent, difficult instances of military campaigning and examine them carefully to discover the full range of military problems they had to address, the chains of causality (theory of victory) employed, and the specific instructions issued by the chief executive to his field commanders. Congress could also direct the war colleges to set aside funding so that surviving senior officials involved in the actual crises might take part in the case studies, adding their own perspectives and insights.

As the experience of the two Bush administrations and the Obama administration have demonstrated, an important subject to be addressed in campaign planning courses is when and how to conclude military operations. George H. W. Bush got it wrong in Iraq in 1991, and George W. Bush got it wrong in Afghanistan in 2001 and in Iraq after 2003. The Obama administration struggles with the issue today. Any future curriculum on campaign planning must include the literature and theory on strategic surrender (e.g., Kecskemeti). Moreover, any future curriculum must devote more time to Phase IV operations. Phase IV is where recent U.S. military efforts have most often gone off the rails. Congress should demand that the service schools and war colleges delve into the details of recent failures, build good case studies on them, and develop the doctrine and force structure requirements (e.g., military governance units) for Phase IV activities in greater detail.

Another step Congress could take would be to require campaign planning to be taught by people who had actually directed it: former combatant command commanders, joint task force commanders, and their operations deputies. The benefit to the students would result from a candid discussion of the campaign planning the retired senior officers engaged in and their notions about how those processes might have been improved.

Costs associated with reforming the curriculum would be minimal, because most of the classes at war colleges undergo nearly constant revision and improvement. Those activities could be refocused to create new case studies of campaign planning. Funding to support guest speakers already exists. The war colleges need only focus their invitations on former senior commanders with actual campaign planning experience. Congress could change the rules for Title 10 instructors (currently civilian academics) to make retired general and flag officers eligible and thus available to teach the strategy and campaign planning courses.

Creating a Counter-Insurgency Method Appropriate to U.S. Forces

If Congress supports restoration of the U.S. military's counter-insurgency and irregular warfare capabilities, it should undertake two efforts: first, enable population control measures like those described earlier in this book; and second, enhance the military's ability to destroy the enemy.

Population control for worst-case circumstances in either counter-insurgency or post-combat stability operations will often require instrumentation of the population and its environs as well as control of critical commodities. Congress should appropriate funding to support instrumentation of foreign cities infrastructure and populations. The exact costs will be dependent on the size of the population and territory that must be controlled, and on the mix and density of technologies required. It would be useful for Congress to provide the DoD with $20 million in overseas contingency operations funds as a starting point for experimentation and field trials. Congress should also direct the U.S. Army Corps of Engineers to serve as the principal organization with responsibility for instrumentation of foreign populations. The army should be directed to raise, train, and equip at least one expeditionary sustainment command specifically for the tasks associated with controlling critical commodities (e.g., food, fuel, potable water) that sustain the populace and render them dependent on U.S. forces until a future transfer of authority to local officials takes place. Such a unit could be established within the Army National Guard and capitalize on the Guard's proven capabilities demonstrated in Iraq and Afghanistan.

When it comes to destroying the enemy, Congress should take another three broad actions. First, it should require the president to make aggressive use of his or her authority to designate identified enemy groups Declared Hostile Forces and sanction the use of military force against them. The United States did as much during the Second World War with Germany and Japan, and it has used this authority on a more limited basis in Iraq and Afghanistan. The practice itself dates back at least to Caesar and the Gallic Wars, where Roman forces made distinctions in their treatment of Gallic tribes based on the quality of their relations with Rome.[2] Second, Congress should grant the president authority to employ coercion, use of force, and other military means against states and other actors that support insurgents and other armed entities posing a threat to U.S. interests or its allies. Such a blanket authority, granted in advance of specific need, would put the forces of anarchy on notice, and authorize U.S. forces to prepare for their future tasks. Finally, Congress should require the DoD to undertake a sweeping review of its rules of engagement and their effect on U.S. and coalition casualties, and overall military effectiveness. The goal should be to make the ROE an effective instrument through which the president as commander-in-chief can control and constrain the use of military force as he or she deems necessary and consistent with the laws of land warfare without imposing unintended consequences on U.S. forces deployed to fight America's wars.

Because many of today's insurgents and militias are often supported by foreign governments, in addition to targeting these forces and destroying them, it is important that the United States punish the states that provide assistance and compel them to cease. The United States should work to broaden its options for punishing such states and compelling them to cease their support. Washington has historically used punitive military strikes and the DoD should be directed to develop a new suite of options appropriate for Iran, Pakistan, and others like them. In this regard, particular attention should be paid to appropriate options and potential redlines when considering punitive action against nuclear-capable states. Russia and China displayed restraint during their 1969 border confrontation. India has displayed restraint in responding to terrorist actions with origins in Pakistan. It is important for the DoD to understand what sort of punitive actions are likely to be sufficiently costly so as to dissuade a state from continuing its support while at the same time causing it to prefer to cease its support rather than pursue escalation of the conflict.

Washington has also employed non-military sanctions and embargoes to deny state adversaries access to their overseas accounts, high technology, rare earths, and other valuable goods. The president should direct the appropriate agencies to develop the cyber capabilities needed to undermine adversarial states' money and banking systems, their telecommunications, utilities, and public transportation systems. Such an enhanced set of options for penalizing states supporting insurgency and insurrection would be an important complement to current punitive options and to boots on the ground military action against the insurgents themselves.

Regime Change and Limited Objective Combat Operations

Recalling the Congressional Research Service report on U.S. military deployments, limited objective combat operations have a checkered history, casting doubts as to their inherent superiority to go-for-broke regime change operations. The noncombatant evacuations and humanitarian and disaster relief operations had better outcomes but are, almost by definition, less demanding than combat. Two guidelines emerge from the study of the United States' recent military operations. First, in considering new military action, policymakers and their military advisors should focus on the chains of causality available to them—that is, establish the threat to U.S. interests that the proposed military operation is to treat, and then map out multiple courses of military action to see which one seems to offer the superior remedy for the threat. Be crystal clear about how the application of force causes the adversary to abandon the policy or behavior that imperils U.S. interests. Congress should insist on a detailed briefing on the chain of causality when it considers future presidential requests for Authorizations for the Use of Military Force.

Second, follow Donald Rumsfeld's example and create a "parade of horribles" to capture the potential problems that could derail the U.S. campaign or cause it to create additional troubles for the United States. Then apply assumption-based planning techniques developed by RAND alumnus James Dewar.[3] This process requires officials to identify the signposts—the hints—that one of the "horribles" may be looming just over the horizon. Signposts, when spotted, should prompt hedging or shaping strategies. Shaping means taking actions to prevent the "horrible" from coming true, while hedging means identifying actions to cope with the "horrible" when it can't be prevented. Such actions could prevent surprises, help senior leaders estimate the risk associated with their plans, and help them manage it.

Real-life military problems rarely lend themselves to straightforward selection of the best course of action (although the military decision-making process tries to provide commanders and their staffs a manageable process for doing so). Generally, the plan with the strongest chain of causality should be preferred, but the president should also pay attention to the length and content of its accompanying "parade of horribles." If the list is long or some of the consequences are truly dire, clearly the commander-in-chief should seek the assurances of his or her military leaders that the risk inherent in the "parade of horribles" is manageable, where what is manageable may be understood in terms of the required hedging and shaping actions. Congress should include these considerations in its legislative actions to shape the future civil-military relationship.

Finally, the scope of a military operation is of little consequence if the armed forces are too small to accomplish it or lack the ability to provide military governance, population security, and stability until the United Nations or some form of civil authority can take the reins. Congress and the military should reject the myth of whole of government solutions and recognize that it will be up to the U.S. armed forces to administer the territory and provide services and security to the populations they conquer. The seeds of those capabilities exist in some specialties such as civil affairs, but they are by themselves inadequate. It is time to recognize the need for military governance units and to create rear area security/military governance formations. The units themselves will be a hard sell to the military services, because creating them will mean devoting service end strength to non-combat formations. Congress should mandate their redevelopment anyway. The units should be part of the Army National Guard and Army Reserve to take advantage of civilian skills possessed by Guardsmen and individual Army Reservists. The units would provide a multitude of civil functions, including those of a mayor's office, to preserve public records, issue identification cards and permits, and hear and address popular grievances. The units would also need a justice section to handle civil and criminal justice cases. Sanitation, water supply, and electricity would require engineers. Public order would call for police, fire, and rescue services. Disarmament and demilitarization of former soldiers and combatants would call for ordnance and explosive ordnance disposal experts, and probably some combat arms, depending on the situation. If military governance units turn out to be around 306 personnel each, the annual operations and maintenance costs should be approximately $21.4 million per unit. These costs could probably be reduced roughly 85 percent by placing them in the Army Reserve and Army National Guard.[4]

The loss of budget and prestige to the Department of State as a result of such actions would be significant, and the State Department would fight back vigorously against any such efforts. However, the department has had over fourteen years to reform the way it operates and to create personnel who can be deployed to dangerous areas involuntarily and has failed to do so. The status quo, which deprives U.S. theater commanders of the full suite of capabilities for governance and order, is no longer acceptable. Congress should use its authority to build the capabilities of U.S. armed forces in the key skills discussed here.

The military instrument is only a useful policy option if the president and senior defense officials are confident that it can accomplish the mission they assign it and that, in doing so, it will create the policy outcomes they were pursuing. This book began with the assertion—now proven—that the problems confronting the military were conceptual and intellectual and embedded in policy and law that require rethinking. The analysis presented demonstrated the problems arising from inconsistent civil-military relations; atrophied campaign planning and execution skills; an inappropriate approach to counter-insurgency, irregular warfare, and stability operations; and an inattention to the chain of causality in designing campaign plans that all contribute to long wars that end badly for the United States and leave the country with a huge bill in terms of sunk costs, reconstitution of damaged equipment, extensive treatment for the wounded, and survivor benefits for the families of those killed.

The Congress and DoD cannot ignore this track record or dismiss it as a series of blunders by earlier administrations. It is time to address the problem head-on. This book has presented a short list of issues that seem to be root causes of trouble in the successful application of military force, and prescribed steps to address them. It is my sincerest hope that Congress will engage on the matter and require the DoD to address these issues.

Notes

1. Estimates based on the FY 2014 OSD Operation and Maintenance Budget, http://comptroller.defense.gov/Portals/45/Documents/defbudget/fy2014/budget_justification/pdf/01_Operation_and_Maintenance/O_M_VOL_1_PART_1/OSD_OP-5.pdf.
2. Julius Caesar, *The Gallic Wars* (Cambridge, MA: Loeb Classic Library, Harvard University Press, 1976), books IV and VII.
3. James A. Dewar, *Assumption-Based Planning: A Tool for Reducing Avoidable Surprises* (Santa Monica, CA: RAND Corporation, 2002).
4. Calculations based on the fiscal year 2015 Operations and Maintenance-Army budget. The Operations and Maintenance budget supports 490,000 soldiers and 106,049 civil servants. Dividing the total budget by total personnel it supports yields a per capita Operations and Maintenance figure of $70,064. The per capita Operations and Maintenance cost multiplied by the military governance unit end strength (306) comes to $21.4 million: a rough estimate of such a unit's cost. See comptroller.defense.gov/Portals/45/Documents/defbudget/fy2015/fy2015_OM_Overview.pdf.

Bibliography

White House Papers

National Security Strategy of the United States, May 2010.

Presidential Decision Directive 25, Clinton Administration Policy on Reforming Multilateral Peace Operations.

9/11 Commission

Final Report of the National Commission on Terrorist Attacks Upon the United States. New York: W. W. Norton, undated.

Department of Defense Documents

Department of Defense casualty website. https://www.dmdc.osd.mil/dcas/pages/casualties_ond.xhtm.

Department of Defense Inspector General's report, DODIG 2015-093, Summary of Lessons Learned—DoD IG Assessment Oversight of "Train, Advise, Assist, and Equip" Operations by U.S. and Coalition Forces in Iraq and Afghanistan. http://www.dodig.mil/pubs/documents/DODIG-2015-093.pdf.

Department of Defense, Quadrennial Defense Review Report, September 30, 2001.

Headquarters, Department of the Army, Field Manual 3-24 Counterinsurgency.

Kretchik, Walter E., Robert F. Baumann, and John T. Fishel. *A Concise History of the U.S. Army in Operation Uphold Democracy*. Fort Leavenworth, KS: U.S. Army Command and General Staff College Press, 1998.

Joint Pub. 1-02.

Joint Pub. 5-0.

Macdonald, Charles B. *World War I: The U.S. Army Overseas*. Office of the Chief of Military History, U.S. Army. http://www.history.army.mil/books/AMH/AMH-18.htm.

National Security Directive 54. http://nsarchive.gwu.edu/NSAEBB/NSAEBB80/.

Office of the Secretary of Defense. FY 2014 OSD Operation and Maintenance Budget. http://comptroller.defense.gov/Portals/45/Documents/defbudget/fy2014/budget_justification/pdf/01_Operation_and_Maintenance/O_M_VOL_1_PART_1/OSD_OP-5.pdf.

Operation Just Cause: the Incursion into Panama. U.S. Army Center for Military History. CMH Pub. No. 70-85-1.

Reed, John A., Jr. *Germany and NATO*. Washington, DC: National Defense University Press, 1987.

Romjue, John L. *From Active Defense to AirLand Battle: The Development of Army Doctrine 1973–1982*. United States Army Training and Doctrine Command Historical Office, June 1984.

U.S. Army. *Chronology of the Occupation 15 August 1945–31 March 1946*. http://www.history.army.mil/documents/8-5/8-5.htm.

U.S. Army. *Operational Law Handbook: International and Operational Law Department*. Charlottesville, VA: Judge Advocate General's Legal Center & School, U.S. Army, 2010. CLAMO@hqda.army.mil.

U.S. Army Center for Military History. http://www.history.army.mil/html/reference/army_flag/ww2_ap.html.

U.S. Army Center for Military History. *The United States Army in Somalia 1992–1994*. CMH Pub. 70-81-1.

U.S. Army Center for Military History. *War in the Persian Gulf: Operations Desert Shield and Desert Storm August 1990–March 1991*. CMH Pub. 70-117-1, 2010.

U.S. Army Combined Arms Center. http://usacac.army.mil/sites/default/files/documents/cace/350-1_CGSCCatalog.pdf.

U.S. Army Field Manual 3-07.1, Security Force Assistance.

U.S. Army War College Student Handbook (37–38). http://dde.carlisle.army.mil/DDE_documents/StudentHandbook_2013.pdf.

U.S. Central Command. CENTCOM commander's Posture Statement before the House Appropriations Committee. March 5, 2015. www.centcom.mil/en/about-centcom-en/commanders-posture-statement-en.

U.S. Special Inspector General for Afghanistan Reconstruction (SIGAR). file:///Users/johnpeters/Desktop/SIGAR:%20Pentagon's%20Economic%20Development%20in%20Afghanistan%20'Accomplished%20Nothing'%20%7C%20Defense%20News%20%7C%20defens.webarchive.

U.S. Strategic Bombing Survey: Japan's Struggle to End the War. July 1, 1946. https://www.trumanlibrary.org/whistlestop/study_collections/bomb/large/documents/pdfs/68.pdf.

State Department Documents

Department of State, Office of the Historian. https://history.state.gov/milestones/1914-1920/haiti.

Department of State, Office of the Historian. https://history.state.gov/milestones/1945-1952/japan-reconstruction.
Department of State, Office of the Historian. https://history.state.gov/milestones/1993-2000/haiti.
U.S. Mission Afghanistan. *U.S. Foreign Assistance for Afghanistan Post Performance Management Plan 2011–2012*, Volume 1.

Congress/Congressional Research Service

Congressional Research Service. *Instances of Use of United States Armed Forces Abroad, 1798–2014*. Report Number R42738, September 15, 2014.
Dale, Catherine. *War in Afghanistan: Strategy, Operations, and Issues for Congress*. Congressional Research Service, R40156. March 2011.
U.S. Congress, Office of Technology Assessment. New Technology for NATO: Implementing Follow-on Forces Attack. June 1987. UNT Digital Library. http://digital.library.unt.edu/ark:/67531/metadc39886/.

United Nations Documents

Dayton Accords, Annex 1-A, Article 1. http://peacemaker.un.org/sites/peacemaker.un.org/files/BA_951121_DaytonAgreement.pdf.
UN Monitoring Verification and Inspection Commission. http://www.un.org/Depts/unmovic/documents/1441.pdf.

Books and Memoirs

Barber, Noel. *The War of the Running Dogs*. London: Cassell, 2004.
Bennett, Huw. *Fighting the Mau Mau: The British Army and Counterinsurgency in the Kenya Emergency*. Cambridge: Cambridge Military Histories, 2012.
Bolger, Daniel P., Lieutenant General (U.S. Army, Ret.). *Why We Lost: A General's Inside Account of the Iraq and Afghanistan Wars*. Boston: Edmond Dolan Books/Houghton-Mifflin Harcourt, 2014.
Boot, Max. *Invisible Armies: An Epic History of Guerrilla Warfare from Ancient Times to the Present*. New York: W. W. Norton, 2013.
Branch, Daniel. *Defeating Mau Mau, Creating Kenya: Counterinsurgency, Civil War, and Decolonization*. Cambridge: Cambridge University Press, 2009.
Brands, H. W. *Woodrow Wilson*. New York: Times Books, Henry Holt, 2003.
Bush, George H. W., and Brent Scowcroft. *A World Transformed*. New York: Vintage, 1999.
Bush, George W. *Decision Points*. New York: Crown, 2010.
Caesar, Julius. *The Gallic Wars*. Cambridge, MA: Loeb Classic Library, Harvard University Press, 1976.
Chin, Truong. *The Resistance Will Win*.

Clark, Christopher. *The Sleepwalkers: How Europe Went to War in 1914*. New York: HarperCollins, 2012.

Clausewitz, Carl von. *On War*. Translated and edited by Michael Howard and Peter Paret. Princeton, NJ: Center of International Studies, Princeton University, 1989.

Coates, John. *Suppressing Insurgency: An Analysis of the Malayan Emergency, 1948–1954*. Boulder, CO: Westview Press, 1992.

Cohen, A. A. *Galula: The Life and Writings of the French Officer Who Defined the Art of Counterinsurgency*. New York: Praeger, 2012.

Cohen, Eliot A. *Supreme Command: Soldiers, Statesmen and Leadership in Wartime*. New York: Free Press, 2002.

Dewar, James A. *Assumption-Based Planning: A Tool for Reducing Avoidable Surprises*. Santa Monica, CA: RAND, 2002.

Dönitz, Karl, Grossadmiral. *Zehn Jahre und Zwanzig Tage*. Translated by R. H. Stevens as *Memoirs: Ten Years and Twenty Days*. Cleveland: World Publishing, 1959.

Feith, Douglas J. *War and Decision: Inside the Pentagon at the Dawn of the War on Terrorism*. New York: Harper, 2008.

Furedi, Frank. *Colonial Wars and the Politics of Third World Nationalism*. London: I.B. Tauris, 1996.

Gall, Carlotta. *The Wrong Enemy: America in Afghanistan, 2001–2014*. Boston: Houghton Mifflin Harcourt, 2014.

Gates, Robert M. *Duty: Memoirs of a Secretary at War*. New York: Alfred A. Knopf, 2014.

Giap, Vo Nguyen. *Peoples' War, Peoples' Army*. New York: Praeger, 1962.

Herspring, Dale R. *The Pentagon and the Presidency: Civil-Military Relations from FDR to George W. Bush*. Lawrence: University Press of Kansas, 2005.

Hindenburg, Paul, and Frederic Appleby Holt. *Out of My Life*. Volume II (1921). Kessinger Publishing's Rare Reprints.

Holbrooke, Richard. *To End a War*. New York: Modern Library, 1999.

Horne, Alistair. *A Savage War of Peace: Algeria 1954–1962*. New York: New York Review Books, 1977.

Huntington, Samuel P. *The Soldier and the State: the Theory and Politics of Civil-Military Relations*. 3rd ed. Cambridge, MA: Belknap Press, Harvard University Press, 1998.

Kahneman, Daniel. *Thinking, Fast and Slow*. New York: Farrar, Straus and Giroux, 2011.

Kecskemeti, Paul. *Strategic Surrender: The Politics of Victory and Defeat*. Santa Monica, CA: RAND Corporation, 1957.

Linn, Brian Michael. *The Echo of Battle: The Army's Way of War*. Cambridge, MA: Harvard University Press, 2007.

Logevall, Fredrik. *Embers of War: The Fall of an Empire and the Making of America's Vietnam*. New York: Random House, 2012.

MacDonogh, Giles. *After the Reich: The Brutal History of the Allied Occupation*. New York: Basic Books, 2007.

Maloba, Wunyabari O. *Mau Mau and Kenya: an Analysis of a Peasant Revolt*. Bloomington: Indiana University Press, 1998.

Milne, David. *Worldmaking: The Art and Science of American Diplomacy*. New York: Farrar, Straus and Giroux, 2015.

Nielson, Suzanne C., and Don M. Snider, eds. *American Civil-Military Relations: The Soldier and the State in a New Era*. Baltimore, MD: Johns Hopkins University Press, 2009.

Overy, Richard. *Russia's War: A History of the Soviet War Effort, 1941–1945*. New York: Penguin Books, 1998.

Paul, Christopher, et al. *Paths to Victory: Lessons from Modern Insurgencies*. Santa Monica, CA: RAND Corporation, 2013.

Paul, Christopher, et al. *Victory Has a Thousand Fathers*. Santa Monica, CA: RAND Corporation, 2010.

Peters, John E., et al. *European Contributions to Operation Allied Force: Implications for Transatlantic Cooperation*. Santa Monica, CA: RAND Corporation, 2001.

Peters, John E., et al. *The U.S. Military: Ready for the New World Order?* Westport, CT: Greenwood Press, 1993.

Rabasa, Angel, John E. Peters, et al. *Ungoverned Territories: Understanding and Reducing Terrorism Risks*. Santa Monica, CA: RAND Corporation, 2007.

Rice, Condoleezza. *No Higher Honor: A Memoir of My Years in Washington*. New York: Crown, 2011.

Rumsfeld, Donald. *Known and Unknown: A Memoir*. New York: Sentinel, 2011.

Schroeter, H. G. *Americanization of the European Economy: A Compact Survey of American Economic Influence in Europe Since the 1800s*. New York: Springer, 2005.

Schwarzkopf, H. Norman, and Peter Petre. *It Doesn't Take a Hero*. New York: Linda Grey Bantam Books, 1992.

Senior, Ian. *Invasion 1914: The Schlieffen Plan to the Battle of the Marne*. New York: Osprey, undated e-book edition.

Shelton, General Hugh. *Without Hesitation: The Odyssey of an American Warrior*. New York: St. Martin's Press, 2010.

Sky, Emma. *The Unraveling: High Hopes and Missed Opportunities in Iraq*. New York: Public Affairs, 2015.

Smith, General Sir Rupert. *Utility of Force: The Art of War in the Modern World*. London: Penguin, 2005.

Stafford, David. *End Game, 1945: The Missing Final Chapter of World War II*. New York: Back Bay Books/Little, Brown, 2007.

Stargardt, Nicholas. *The German War: A Nation Under Arms 1939–1945*. New York: Basic Books, 2015.

Windrow, Martin. *The Last Valley: Dien Bien Phu and the French Defeat in Vietnam*. Cambridge, MA: Da Capo Press, 2004.

Journals and Magazines

Brennan, Rick. "Withdrawal Symptoms: The Bungling of the Iraq Exit." *Foreign Affairs* (November/December 2014).

Chorley, Matt, and Leon Watson. "Mau Mau Fighters Raped, Castrated and Beaten in Kenya's Uprising against Britain get £14 million Compensation But No Apology." *Daily Mail*, June 5, 2013. http://www.dailymail.co.uk/news/article-2336418/Mau-Mau-fighters-raped-castrated-beaten-Kenyas-uprising-Britain-14million-compensation-apology.html.

Clad, James. "Wasting the Golden Hour in America's Iraq Meltdown." *National Interest* (July-August 2013).

Daalder, Ivo. "Decision to Intervene: How the War in Bosnia Ended." Brookings Institution. http://www.brookings.edu/research/articles/1998/12/balkans-daalder.

Deery, Phillip. "Malaya, 1948: Britain's Asian Cold War?" *Journal of Cold War Studies* 9, no. 1 (2007): 29–54.

deGrandpre, Andrew, and Andrew Tilghman. "Iran Linked to Deaths of 500 U.S. Troops in Iraq, Afghanistan." *Military Times*, July 15, 2015. http://www.militarytimes.com/story/military/capitol-hill/2015/07/14/iran-linked-to-deaths-of-500-us-troops-in-iraq-afghanistan/30131097/.

Fair, C. Christine, and Sumit Ganguly. "An Unworthy Ally." *Foreign Affairs* (September–October 2015).

Fallows, James. "The Tragedy of the American Military." *Atlantic*. http://www.theatlantic.com/magazine/archive/2015/01/the-tragedy-of-the-american-military/383516/.

Foreign Affairs (July-August 2014) (entire issue).

Goldstein, Lyle J. "General John Shalikashvili and the Civil-Military Relations of Peacekeeping." *Armed Forces and Society* 26, no. 3 (Spring 2000).

Jeffrey, James F. "Why Counterinsurgency Doesn't Work." *Foreign Affairs* (March/April 2015).

Knowles, David. "Rumsfeld: George W. Bush Was Wrong About Iraq." *Bloomberg Politics*, June 8, 2015. http://www.bloomberg.com/politics/articles/2015-06-08/donald-rumsfeld-george-w-bush-was-wrong-about-iraq.

Koehl, Stuart L., and Stephen P. Glick. "Why the Rescue Failed: There Was More to Operation Eagle Claw's Failure in the Desert of Iran than Jimmy Told Us." *American Spectator*, July 1980. http://spectator.org/articles/34807/why-rescue-failed.

Kohn, Richard H. "Out of Control: the Crisis in Civil-Military Relations." *National Interest* (Spring 1993/1994).

Moore, Lieutenant Colonel Charles P. "What's the Matter with Being a Strategist (Now)?" *Parameters* (Winter 2009–2010).

Nix, Commander Dayne E. "American Civil-Military Relations: Samuel P. Huntington and the Political Dimensions of Military Professionalism." *Naval War College Review* 65, no. 2 (Spring 2012).

Panda, Ankit. "Why China Snubbed India on a Pakistan-based Terrorist at the UN." *The Diplomat*, June 25, 2015. http://thediplomat.com/2015/06/why-china-snubbed-india-on-a-pakistan-based-terrorist-at-the-un/.

Peters, John E. "Evaluating FOFA as a Deterrent." *RUSI Journal* (December 1987): 39–46.

Peters, John E. "A Potential Vulnerability to Precision-Strike Warfare?" *Orbis* (Summer 2004).

Quinlivan, James T. "Force Requirements for Stability Operations." *Parameters* (1996).

Ramakrishna, Kumar. "'Telling the Simple People the Truth': The Role of Strategic Propaganda in the Malayan Emergency." *Journal of the Malaysian Branch of the Royal Asiatic Society* 75, no. 1 (2002): 49–68.

Ramakrishna, Kumar. "'Bribing the Reds to Give Up': Rewards Policy in the Malayan Emergency." *War in History* 9, no. 3 (2002): 332–53.

Newspapers and Information Services

Alexander, Harriet. "US Special Forces Evacuated from Yemen as Security Situation Deteriorates." www.telegraph.co.uk/news/worldnews/middleeast/yemen/11487729/US-special-forces-evacuate-from-yemen-as-security-situation-deteriorates.html.

Ayres, Alyssa. Council on Foreign Relations Policy Innovation Memorandum 53. "Why the United States Should Work with India to Stabilize Afghanistan." http://www.cfr.org/afghanistan/why-united-states-should-work-india-stabilize-afghanistan/p36414.

Balkan Witness. http://balkanwitness.glypx.com/racak.htm.

Brezhnev Doctrine. http://www.britannica.com/EBchecked/topic/79119/Brezhnev-Doctrine.

Goldwater-Nichols DoD Reorganization Act of 1986. http://history.defense.gov/Portals/70/Documents/dod_reforms/Goldwater-NicholsDoDReordAct1986.pdf.

Gordon, Michael R. "Failure of Haiti Operation Backs Initial Pentagon Skepticism." *New York Times*, October 15, 1993. http://www.nytimes.com/1993/10/15/world/failure-of-haiti-operation-backs-initial-pentagon-skepticism.html.

Grant, Wil. "Has Panama Weaned Itself Off Drugs and Cleaned Up?" August 3, 2014. http://www.bbc.com/news/world-latin-america-27945558.

"Iraqi Detainees at Guantanamo Bay." *Wikipedia*. https://en.wikipedia.org/wiki/Iraqi_detainees_at_Guantanamo_Bay.

Ledeen, Gabriel. "Re-creating Anbar's Awakening." *Huffington Post*, April 27, 2009. http://www.huffingtonpost.com/gabriel-ledeen/re--creating-anbars-awake_b_179589.html.

LTG Thomas Montgomery interview. http://www.pbs.org/wgbh/pages/frontline/shows/ambush/interviews/montgomery.html.

NATO. "The Future Tasks of the Alliance" (Harmel Report). December 13, 1967. http://www.nato.int/cps/en/natohq/official_texts_26700.htm.

New York Times. "U.S. Military Chief Is Replaced In the Central American Region." http://www.nytimes.com/1989/07/21/world/us-military-chief-is-replaced-in-the-central-american-region.html.

PBS *Frontline*. "Biography of Admiral Leighton Smith." http://www.pbs.org/wgbh/pages/frontline/shows/military/guys/sbio.html.

PBS *Frontline*. "The Rules of Engagement." http://www.pbs.org/wgbh/pages/frontline/haditha/themes/roe.html.

Perry, Mark. "U.S. Generals Think Saudi Strikes in Yemen a Bad Idea." April 17, 2015. www.america.aljazeera.com/articles/2015/4/17/US-generals-think-saudi-strikes-in-yemen-a-bad-idea.html.

Référendum sur l'autodétermination en Algérie. http://mjp.univ-perp.fr/france/ref1961.htm.

Scarborough, Rowan. "Shades of Vietnam: Spike in U.S. Troop Deaths Tied to Stricter Rules of Engagement." *Washington Times*, December 5, 2013. http://www.washingtontimes.com/news/2013/dec/5/increase-in-battlefield-deaths-linked-to-new-rules/?page=all.

Stockholm International Peace Research Institute. http://www.sipri.org/yearbook/2015/downloadable-files/sipri-yearbook-2015-summary-pdf.

Tan, Michelle. "Hundreds Support Family of Fallen Green Beret; Lawmakers Seek Answers to Deadly Battle." *Military Times*, January 7, 2016. http://www.armytimes.com/story/military/2016/01/07/hundreds-support-family-fallen-green-beret-lawmakers-seek-answers-deadly-battle/78421900/.

Taylor, Adam. "Four Months Ago, Obama Called Yemen's War on Terror a Success. Now the Yemeni Government May Fall." *Washington Post*, January 20, 2015. www.washingtonpost.com/blogs/worldviews/wp/2015/01/20.

Truth and Action. "U.S. Troops Fed Up with 'Rules of Engagement' That Strengthen the Enemy." http://www.truthandaction.org/us-troops-fed-rules-engagement-strenghten-enemy/.

Wall Street Journal. "Leadership Lessons from the Generals." December 9, 2014.

Wall Street Journal. "Civilians in Crosshairs Slow Troops." http://www.wsj.com/articles/SB10001424052748704751304575079741450028512.

War Chronicle. http://warchronicle.com/numbers/WWII/deaths.htm.

War on the Rocks. http://warontherocks.com/category/strategic-outpost/.

Index

Lightning Source UK Ltd.
Milton Keynes UK
UKOW04n0319161217
314508UK00011B/643/P

9 781442 274716